Data and Error Analysis

Data and Error Analysis

Second Edition

William Lichten
Yale University

Prentice Hall, *Upper Saddle River, New Jersey 07458*

Library of Congress Cataloging-in-Publication Data
Lichten, William
Data and error analysis
William Lichten. — 2nd ed.
p. cm.
Includes bibliographical references and index.
ISBN 0-13-368580-2
1. Statistics. 2. Error analysis (Mathematics) I. Title.
QA276. 12.L53 1999
519.5—dc21 98-33710
CIP

EXECUTIVE EDITOR: Alison Reeves
EXECUTIVE MANAGING EDITOR: Kathleen Schiaparelli
ASSISTANT MANAGING EDITOR: Lisa Kinne
ART DIRECTOR: Jayne Conte
MANUFACTURING MANAGER: Trudy Pisciotti
COVER DESIGNER: Bruce Kenselaar
PRODUCTION SUPERVISION/COMPOSITION: Monotype Composition Editorial Services

Printed in the United States of America
10 9 8 7 6 5 4 3 2

ISBN 0-13-368580-2

PRENTICE-HALL INTERNATIONAL (UK) LIMITED, *LONDON*
PRENTICE-HALL OF AUSTRALIA PTY. LIMITED, *SYDNEY*
PRENTICE-HALL CANADA INC., *TORONTO*
PRENTICE-HALL HISPANOAMERICANA, S.A., *MEXICO*
PRENTICE-HALL OF INDIA PRIVATE LIMITED, *NEW DELHI*
PRENTICE-HALL OF JAPAN, INC., *TOKYO*
PEARSON EDUCATION ASIA PTE. LTD., *SINGAPORE*
EDITORA PRENTICE-HALL DO BRASIL, LTDA., *RIO DE JANEIRO*

CONTENTS

To Susan

PREFACE AND INTRODUCTION

This is the second edition of a book to help science students process their data, without lengthy and boring computations. It gets to the point without long discussions and involved derivations. It gives simple, handy rules for estimating errors, both by graphical and analytic methods. It takes advantage of new data processing software and hardware, which have been tried out by students in the author's classes. In particular, it estimates errors in correctly weighted regression: linear, log, exponential, exponential with counts, and power law fits.

This edition also adds a new section on power law relations (log-log plots); eliminates lengthy, obsolete, algebraic and graphical analyses; and has a reorganized, consolidated, and expanded section on probability distributions. The latter includes a discussion of student's t-distribution, which is the correct one for small sample errors. The entire approach to errors for numerical fits has been improved to make a clear distinction between *a priori* estimates and those based on data scatter.

The second half of the book, the programs for calculators and computers, is totally new. Two sets of programs on CD-ROM practice painless data analysis. The first uses Excel™, the most widely-used spreadsheet (IBM or Mac format). It takes in the data and, in seconds, does all of the drudgery of plotting, statistics, and data fits. The second is a cT program which does error analysis automatically, thereby bypassing perhaps the sorest point for students in laboratory courses. The computer programs were written by Dr. Donald Shirer, the director of the Yale Physics Department undergraduate laboratories. Also, it has BASIC and Pascal programs in a form which is compatible with almost all computers which have these languages.

The material in this book should be useful for physics and any other subject that uses exact methods of measurement, such as chemistry, biophysics, or geology.

The emphasis in this book is on "learning by doing." Thus, it gives useful algorithms and programs, rather than theoretical derivations. This approach comes from the author's experience of the needs of students confronted with error analysis and statistics for the first time in their lives. For those students who wish to delve more deeply into the theory, references are given to the standard texts and monographs in this field.

BUYING AND USING YOUR CALCULATOR

This book assumes that you, the reader, have at least a two-variable, scientific calculator (one that has σ_x, σ_y, a and b keys). A list of calculators for which this book provides programs is in Appendix B.

Priority: the order of calculating. Every calculator has a language that you must learn before you start punching keys. Most calculators will do multiplication and division first, then addition and subtraction: for example, $2 + 3 \times 4 = 2 + 12 = 14$. You'll need to know the priority of your calculator's functions, like 1^{-}, y^x., sin, etc., along with $+, -, \times, /$.

Round-off. The large number of digits in the calculator display is sometimes a problem, which we solve by rounding off. All calculators give an answer to only so many digits. This sometimes leads to strange results. For example take 1/9, multiply it by 9, then subtract 1. The answer is zero, right? Try it on your calculator and notice that you may get 0.00000001, 1E-10, or some such strange answer. The reason for these answers is the limited number of digits in the calculator. This is called "round-off". It occasionally can lead to problems in statistical calculations; we'll watch out for this when we get to that subject.

Memory. Any calculator remembers the last number in its display, which you can retrieve with the "answer" key, after you key in the next number or operation. Scientific calculators have quite a few additional memories (Σ^+, *DATA*, *A*, *B*, *C*, etc., or *LISTS* for graphic calculators) for storing data, which often come in handy. *Programmable* calculators grant you the labor-saving ability to store a whole sequence of key strokes.

Computers vs. calculators. Calculators can solve quite complicated data processing problems. However, computers, with their spreadsheet programs, graphing, and printing capabilities are unsurpassed for ease and versatility in the analysis of data. An ideal solution is to have a small, inexpensive pocket calculator for the simpler, everyday, calculations and a computer with a spreadsheet program to handle more complex work and to prepare and to print tables and graphs. The author is unimpressed with the quality of calculator graphs.

"Garbage." A common problem with memories is that they hold "garbage" you don't want in your calculations. Since most calculators have "continuous memory", data isn't lost even if you turn the calculator off. Some calculators automatically clear memory on turn-on or when you run a program (see next paragraph for explanation of this term). Remember to clear the memory before every new calculation. This is essential with statistical data.

A program, what is it? Regardless of the type of calculator you use, you'll need a *program* to solve your more complex data processing problems. Without a program, you'll find yourself filling sheets of paper with scribbled numbers and finding a different answer every time you do a calculation.

A *program* is simply writing down in skeleton form all the keystrokes you'll need to solve a problem. You follow this list as you do a calculation. The only things you change are the data, the numbers from your experiment which you feed into your program. *Programmable calculators* and *computers* have the advantage of being able to learn this program from you. You teach the program by keying it in. The calculator or computer then follows this list faithfully every time you need to repeat a calculation.

Fairly complicated calculations, like linear regression (fitting a straight line to a set of points) are easy to do with a programmable calculator. Even if you have a non-programmable calculator which requires keying in every step, the programs in this book are surprisingly easy and fast for solving this type of problem, once you get the hang of it.

Graphs; are they any use? Even in the world of sophisticated calculators and computers, graphs are useful for seeing what's going on in your data. You can have the best of both worlds with a computer with a spreadsheet program which both plots data and processes it. The Excel™ programs enclosed with this book require only the raw data; the computer does all the rest.

Pre-programmed calculators. All scientific calculators have labor-saving statistical calculations wired in to their brains. Using these programs blindly can lead to trouble. One feature of this book is that it helps to avoid this trouble by explaining how these programs work.

What is an algorithm? An algorithm is a set of steps that you follow to solve a problem. The simplest algorithm is a formula, like $y = x^2$. More complex algorithms can contain a chain of such steps. Every program, whether you write it or whether it is "hard-wired" into your calculator, follows an algorithm. In any case, this book gives the algorithms you'll need for your data processing in the laboratory.

HOW TO USE THIS BOOK

The early chapters of this book present the simple principles that are useful to all. The later chapters present a wide variety of methods and programs for solving problems. No one will want to learn *all* these methods. The key to using these chapters is to be *selective*. Pick out a method that is agreeable to you, learn how it works and you will be able to use it to solve many of your data processing problems. If your method does not work in a new situation, you always can come back to this book and try out an alternative solution.

The author thanks two of his students, Jacob Wouden and Michelle Denburg, who are the "Jack and Jill" who performed the experiments on tungsten filaments in Chapter 4.

William Lichten

Data and Error Analysis

1

MEASUREMENT AND ERRORS

1-1 EXACT AND APPROXIMATE STATEMENTS IN SCIENCE

The statement "$2 + 2 = 4$" is *exact*. Such an equation is a matter of the logic of simple arithmetic. It is either true or false. Likewise, "My dog has four legs and a tail" is also exact. To see if this is true is merely a matter of counting.

Is it correct for you to say, "My new pencil is exactly 192 millimeters long"? Let's suppose you measured the length of your pencil with a ruler. If you used a more exact device, such as a pair of vernier calipers, you might say, "Correction! My pencil is 192.16 millimeters long." Your first measurement is good to the nearest millimeter; your second is good to the nearest 0.01 mm. We say that both values are inexact or approximate; both are subject to *errors*. This chapter discusses errors and what they are.

Approximate statements play an important role in science. Most laboratory measurements are approximate. Therefore, the laws of science, which are based on measurements, are only approximately correct. Any law of science might be proved incorrect or might have to be changed to fit new measurements that are more precise.

1-2 ERRORS

Mistakes versus errors. Error has a special meaning in science. Error has a different meaning from mistake. Mistakes, such as measuring a 32-cm-long object to be 42 cm, can be avoided. Errors cannot be avoided, even by the most careful measurements. A public opinion poll to predict an election would make a mistake if it interviewed only Princeton University students, who do not necessarily vote the same as the general population. Instead, the pollster should interview a carefully selected sample of, say, 1000 persons of different incomes, educational levels, races, states, and so on. But even the most carefully chosen sample is subject to errors, which arise from the chancy nature of sampling. Thus the pollster may say, "My results are subject to a 3% margin of error."

Precision Versus Accuracy: Random and Systematic Errors

Let's go back to the example of the pencil. Suppose everyone in the class uses the same ruler, measures the pencil to the nearest millimeter, and all agree it is 192 mm long. All say that it couldn't be either 191 or 193 mm long. We say that the class has measured the length of the ruler to a *precision* of 1 mm. *Precision* is the reliability or repeatability of a measurement.

Suppose that the instructor now points out, "You all have made the same mistake. You lined up one end of the pencil and one end of the ruler together. The end of the ruler is worn badly; it doesn't begin at zero. Try to remeasure the pencil by putting it in the middle of the ruler. Then find the position of both ends." (See Table 1-1.) "Subtract one value from the other to find the length." Now the class finds that the pencil is 187 mm long! How can this be?

Both measurements are equally precise. The second one is more *accurate* than the first, because a *systematic* error (caused by the worn end of the ruler) is no longer there. Another example of a systematic error occurs in weighing with a balance with the zero (reading with no load) improperly adjusted. A *systematic error* is an effect that changes all measurements by the same amount or by the same percentage.

The class's experience with the ruler is a mirror of the history of science. Systematic errors have often hidden unsuspected in measurements. The only way to eliminate systematic errors is to look carefully for them and to understand well the nature of the experiment or measurement.

Random errors. Can we avoid them? Let's return to the example of the class measurement of the length of a pencil; when measuring to the nearest millimeter, everyone got the same value. Let's try to push the precision further and ask each person to measure to the nearest *tenth* of a millimeter. Now disagreements appear. We find different values: 186.7, 187.0, 187.3 mm, as shown in Table 1-1.

Is someone making a mistake? No, even the most careful and skillful person will come up with values that vary by one- or two-tenths of a millimeter. Now we are at the limit of measurement by use of the naked eye and rulers. The unavoidable change in successive measurements, due to small irregularities in the ruler, difficulty in estimating precisely, and the like, is called a *random error*, or *error* for short.

Your best estimate, the arithmetic mean. At this point, you've been careful not to make any mistakes, you've avoided all systematic errors, and you've narrowed your uncertainty to the random error of measurement. What's next?

Common sense tells you to take the average of several measurements, called the *arithmetic mean* or *mean*. The algebraic expression for the average $\overline{X}$ of N numbers is

$$\overline{X} = \frac{\text{sum}}{N} = \frac{x_1 + x_2 + \cdots + x_N}{N} = \frac{\Sigma x}{N} \qquad (1\text{-}1)$$

TABLE 1-1 Measurement of the Length of a Pencil.

Left End, L (cm)	Right End, R (cm)	Length (cm) $= R - L$	Deviation from Mean
10.16	28.83	18.67	−0.03
15.87	34.57	18.70	0.00
20.22	38.95	18.73	+0.03

Sum = 56.10 cm; $N = 3$

Average = sum/N = 18.70 cm

Average deviation = (0.03 + 0.0 + 0.03)/3 = 0.02 cm

What is your error? There are two different methods of finding your error. The first is to *estimate* it in advance (also called the *a priori* method). The second, the *scatter* method, uses the random spread in the measurements. The sections that follow describe how to make such estimates.

Estimation in advance (*a priori* method). In the example of the pencil where every reading was 187 mm, *no random error was present.* When all measurements are the same, use the *least count* (the smallest division on your scale) for your error estimate. In the case of the pencil, the quoted length is 187 $\pm$ 1 mm (or more confidently, 187 $\pm \frac{1}{2}$ mm).

Digital measurements. Digital equipment, such as scales, timers, frequency counters, etc., often read exactly the same value, with no scatter in the data. In such a case, the equipment manual often gives the error as the least count (one digit) plus the systematic error, usually given in percent (see Section 1-4).

Counting and polling. In taking a public opinion poll, in other measurements of percentages of occurrence, or in following radioactive decay, one counts events. (Do not confuse the use of the word "count" here with "least count.") Chapter 2 (binomial distribution) and Chapter 5 (Poisson distribution) give theoretical formulas for estimating counting errors.

Error analysis. Often one measures several quantities and puts the separate results into a formula to calculate a derived quantity. For example, the density ρ of a liquid of mass M and volume V is given by the formula $\rho = M/V$. Chapter 3 discusses how to put together the separate errors in M and V to estimate the error in ρ.

Data scatter. Recall the example on this page, where scatter appeared in measurements made to the nearest tenth of a millimeter. We can use this scatter to find the random error of measurement. A handy measure is the *average* *deviation from the mean*, sometimes shortened to *average deviation*. You can get this by finding the difference between each measurement and the mean and then taking the average. (You count all deviations as positive for this calculation.) An example of these calculations is given in Table 1-1.

The final result is 18.70(2) cm = 18.70 ± 0.02 cm. Note two ways of showing the error: ± precedes the error, or parentheses show the error in the last place. We will see later that, if you take three measurements, the average deviation is a fairly good estimate of the error of your measurement.

Let's go over this again: *Take three measurements. Take the average as your best estimate of the true value. Take the average deviation as an estimate of the error of measurement.* This is a good rule of thumb that has several advantages. It's simple. It's easy to do the calculations (see Table 1-1); most of the time you can do them in your head or on a very small piece of paper.

What do you do with your error? Ever since Galileo dropped weights from the Leaning Tower of Pisa, the history of science is one of laws being broken and new laws taking the place of old. In physics labs the game is to reenact history by testing laws of nature. You will make a measurement, find its *deviation from the accepted value*, and compare this *deviation* with your *error*. If the deviation exceeds the error by a safe margin, you have to come up with an explanation. Maybe you have discovered a new law of physics!

1-3 SIGNIFICANT FIGURES

Rounding off to the right number of significant figures. Using calculators causes a common problem: What do we do with the long string of digits in the display? Keep them all? Throw out some? If so, how many?

The answer to these questions is: keep only the *significant digits;* round off to the correct number of *significant figures.* Knowing the error of your measurement tells you how many significant figures there are in your result. Thus, you never give a result like 23.343 g when you only can weigh to 0.1 g. The correct value is 23.3(1) g. *The number of significant figures is the number of digits needed to state the result of a measurement*, or a calculation based on that measurement, *without losing any precision*. Thus a measurement of 10.05(1) cm (or 10.05 ± 0.01 cm) has four significant figures.

Changing the units of a measurement usually does not change the number of significant figures. Thus 10.05(1) cm = 100.5(1)mm = 0.01005(1) m, all of which have four significant figures. Multiplying or dividing a number by any exact constant, such as 10, 2, 5, or 4, usually doesn't change the number of significant figures.

When adding or subtracting two or more numbers, round the result to the last decimal place of the least precise result. Thus 10.01 cm + 352.2 cm +

0.0062 cm = 362.2161 cm, which we round to 362.2 cm. We round it to the first decimal place, since it is the last decimal of the least precise number 352.2 cm.

When multiplying or dividing two numbers, round to the smaller number of significant figures. Thus a rectangle of length 63.52 cm (four significant figures) and height 3.17 cm (three significant figures) has an area of $A = L \times H = 63.52\ \text{cm} \times 3.17\ \text{cm} = 201.3584\ \text{cm}^2$, which we round off to three significant figures: 201 cm^2.

When in doubt, keep an extra significant figure (e.g., from the last example use 201.4 cm^2), but *never* write down all the figures from your calculator. Most of these digits are "garbage" and only serve to annoy your laboratory instructor or any other reader of your report.

Another rule to remember is to add a significant digit when the first digit is a 1. Thus $3 \times 0.34 = 1.02$, not 1.0. (In case you wondered why the word "usually" crept into the rules for the number of significant figures, this rule causes the exception. In the example given here, multiplying by 3 adds a significant figure.)

One last question: How many significant figures should we write down when we quote an error? The answer to this question is usually one. The reason for this is that errors seldom are known precisely, since we can only estimate our errors. (If we knew *exactly* what our error was, we'd know exactly what we measured and there wouldn't be any error!)

The whole business of significant figures is a simplification of the subject of *error analysis*, which we'll discuss later. When you understand error analysis, you'll be able to see where the significant figures rules come from.

1-4 RELATIVE AND PERCENTAGE ERRORS

Absolute and relative errors. In this chapter an error has been an error: it has been so many millimeters or so many grams. (Sometimes the name *absolute error* is used in this connection.) Yet things are not that simple. If an astronomer gave the distance to the moon to the nearest meter, we would consider it a breathtaking triumph of modern space science. But if you wanted to order a mechanical pencil lead or drafting pen tip over the telephone, you would need to know the diameter to a fraction of a millimeter. *Errors often are a relative matter.* In scientific measurements, it often is meaningful to express errors in fractions or percents.

Relative errors. The *relative of fractional error* is expressed as a fraction:

$$\text{Relative error in a quantity} = \frac{\text{error}}{\text{measured quantity}} \tag{1-2}$$

Likewise, the *percentage error* is 100 times the fractional error:

$$\text{Percentage error in a quantity} = \frac{100 \times \text{error}}{\text{measured quantity}} \tag{1-3}$$

1-5 STANDARDS AND THE BOTTOM LINE

Even the best error estimates are fallible: the history of physics is full of underestimated errors. Whether your errors come from advance estimates or from data scatter, the safest policy is to use a "bottom line" error estimate that makes the fewest assumptions possible. Chemical and medical analysis laboratories do not rely exclusively on error estimates, but check their results directly by routinely slipping accurate, *standard* samples as quality controls into their test batches.

You can do likewise. For example, if you are measuring the density of a liquid by use of Archimedes' principle, try running a water sample (known density 1 g/cm^3). If you fail to get the accepted value with your standard, you have two options to pursue. One is to use the standard to correct for your systematic errors. The other is to make a bottom line error estimate. (See Example 1-3.)

Example 1-1 Errors in Measuring with a Meter Stick

Discuss the two sources of error of measurement with a meter stick. The first occurs because 1 mm is worn off the zero end of the stick. The second is due to a uniform shrinking of the meter stick over its entire length by 1 mm. In particular, calculate the different types of errors for measuring two objects: one is 0.999 m long; the second is 10 cm long.

Solution. The worn end of the ruler causes the same error. No matter what the length of the object, it will appear to be 1 mm longer than its true value. On the other hand, the shrinkage of the meter stick causes the same *fractional* or *percentage* error.

For the long object, we first note that, in both cases, the meter stick is actually 999 mm long. In both cases, the 999-mm-long object would appear to be 1 m long and the error in measuring the length would be 1 mm. The fractional error would be, by Eq. (1-2), 1 mm/999 mm $\approx \frac{1}{1000}$. The percentage error would be, by Eq. (1-3), $100 \times \frac{1}{999} \approx 0.1\%$.

For the short object, the errors for the two sources are different. The worn end causes an error of 1 mm, a fractional error by Eq. (1-2) of 1 mm/100 mm $= 0.01$, and a percentage error, by Eq. (1-3), of $100 \times 0.01 = 1\%$. The shrinkage of a 10 cm $= 0.1$ m length of the ruler is only one-tenth of the shrinkage of 1 m. Thus the error is 0.1 mm. The fractional error is, by Eq. (1-2), 0.1 mm/10 cm $= \frac{0.1}{100} = 0.001$, the same as for the long object. The percentage error by Eq. (1-3) is again $100 \times 0.001 = 0.1\%$. *The uniform shrinkage or expansion of a meter stick or any other scale causes the same fractional or percentage error.*

Example 1-2 Weighing a Sample

An electronic balance has a digital scale and is rated at $\pm 1\%$ or ± 3 least counts, where the least count is 0.1 g. A sample is weighed in a dish that measures 100.0 g empty and 110.0 g loaded. What is the sample mass?

Solution. We use the specifications to estimate errors. The weighing error is the sum of the systematic error, 1% *of the 10-g sample*, and the random error of ± 3 counts, which add to 0.4 g or 4% of the sample mass. The mass of the dish does not affect the final result. One can "tare" the balance (reset the balance zero with the dish alone) and offset the dish mass without error. The remaining error is due to sample alone.

Example 1-3 Jane Looks at the Bottom Line

Jane uses Archimedes' principle to measure the density ρ of a fluid. She uses the formula $\rho = M/V$, where a body of volume V "loses" weight corresponding to a mass M, when she lowers the body in the fluid. She estimates the volume of her test body to be 10 cm^3 and finds a weight loss of 12.6 ± 0.1 g in an unknown fluid. She then checks her method by using water ($\rho = 1.0$ g/cm^3) and finds a weight loss of 9.0 ± 0.1 g, instead of the expected value of 10 g. What should she do?

Solution. Jane first decides whether or not the deviation of measured volume is outside her error. Jane notes that her measured mass is between the limits of $12.6 + 0.1 = 12.7$ g and $12.6 - 0.1 = 12.5$ g, which gives a density between 12.7/10 cm^3 = 1.27 g/cm^3 and 12.5 g/10 cm^3 = 1.25 g/cm^3. Thus her quoted density is 1.26 ± 0.01 g/cm^3, an error of 0.01 g/cm^3. A similar calculation gives Jane's standard's density to be 0.90 ± 0.01 g/cm^3). The deviation of this from the accepted value of 1.0 g/cm^3 is 0.10 g/cm^3, which is much more than her error.

She can play safe by quoting a bottom line percentage error equal to the 10% difference between her measured and accepted values for water, her standard liquid. Her final value is 1.26 ± 0.13 g/cm^3. Alternatively, she can correct her test body's volume to agree with its weight loss in water, $V = 0.90 \pm 0.01$ cm^3. Her corrected density then becomes $\rho = (1.26 \div 0.90)$ g/cm^3 = 1.40 g/cm^3. (Chapter 3, Section 3-3, pp. 35–37 gives the procedure for calculating the error. It is 0.019 g/cm^3.)

Summary

Aside from mistakes and blunders, errors of measurement fall into two types: *systematic* and *random*. Systematic errors usually have the same sign and magnitude for a particular measurement. Systematic errors can be eliminated or corrected by careful attention to the details of an experiment. Random errors occur with either sign and with a variable magnitude. Random errors can't be eliminated, but can be minimized by taking the average of several observations of the same quantity. A simple rule of thumb to estimate the error of the mean of three measurements is to take the average deviation from the mean.

One can estimate errors either in advance or from the data scatter. However, the safest policy is to use the "bottom line" by repeating the measurement on a known standard. Any deviation between the measurement and the standard gives a correction or bottom line error.

Significant figures are those digits needed to state a result without losing any precision. When adding or subtracting numbers, round to the last decimal place of the least precise number. When multiplying or dividing two numbers, round to the smaller number of significant figures. Add a significant digit when the first digit is a 1. Quote one significant digit in errors.

The relative error in a quantity is the ratio of error/quantity; the percentage error is 100 times the relative error.

Problems

Problems marked with an asterisk (*) are more difficult.

1-1. Which statement is exact? Which is approximate? (a) A kilogram equals 1000 g. (b) The population of the United States on April 1, 1990, was 248,718,301. (c) 1 in. = 2.54 cm. (d) The oxygen atom contains eight electrons. (e) My meter stick is divided into 1000 divisions. (f) Each division on my meter stick is 1 mm long.

1-2. Your weight on a bathroom scale is 135 lb. When you get off, the scale reads 2 lb. How much do you weigh?

1-3. Three millimeters are worn off the end of your meter stick. You place the worn end on your desk top and measure the height of the first shelf of a bookcase to be 41.6 cm. The second shelf measures 72.0 cm above the desk. (a) Give the correct height of the first shelf. (b) Give the height of the second shelf above the first. (c) Is it necessary to know how much is worn off the end to get the answer to part (a)? (d) Answer the same question for part (b).

1-4. You weigh your dog by standing on the scales with him in your arms. The reading is $201\frac{1}{2}$ lb. By yourself you weigh 170 lb. (a) What does your dog weigh? After you get off the scales, you find a zero weight reading of 6 lb. (b) What is your correct weight? (c) Is your estimate of the dog's weight in part (a) affected by the zero-weight reading? Why?

1-5. Each of these expressions gives a measurement and its error in the ± or last-digit form. Write each expression in the other form. (a) 12.1 ± 0.5 cm, (b) 234.67(3) g, (c) 983.5(12) g, (d) 56.23 ± 0.17 cm.

1-6. An automobile speedometer has a range of 0 to 100 mph. The error in any reading is 2% of the maximum speed. (a) What is the error at 100 mph? (b) At 50 mph? (c) What is the percentage error at 50 mph? (d) What is the percentage error at 5 mph? (e) Where is the percentage error smallest, and what is it there?

1-7. A digital balance reads the mass M of a body to many figures, not all of which are significant. Round off the results and give the error, with the correct number of significant figures for both. The percentage error is P. (a) $M = 261.65$ g, $P = 0.1\%$. (b) $M = 63.29$ g, $P = 0.01\%$. (c) $M = 1029.72$ g, $P = 1\%$.

1-8. You have a summer job weighing babies in a hospital. The smallest division ("the least count") on the scales is 100 g. (a) What is the maximum possible error of the babies' mass (assuming you make no mistakes)? *(b) On the average, what will be your error (again assuming no mistakes)?

1-9. John measures his lab partner's height to be 176.35 cm, with an error of 0.21 cm. (a) Round the height and error to the correct number of significant figures. (b) Give the same answer in meters, rounded correctly. (c) Repeat with the answer in decimal inches (10.125 not $10\frac{1}{8}$), where 1 cm = 0.3937 in.

1-10. Mary uses a stopwatch to measure the period of the grandmother clock in the college dean's office. Her results are 0.6, 0.7, and 0.6 sec. (a) What is her mean value? (b) Average deviation? (c) Best estimate of the error of measurement? (d) Her fractional error? (e) Her percentage error? (f) Write down her measurement and express its error both ways (± and error in last digit).

1-11. A student uses an electronic stop watch to measure the effect on the period of a pendulum of finite amplitude. The period with a small amplitude of 1° is 1.5000 sec. The period with an amplitude of 30° is 1.5260 sec. The random errors of measurement are 1% ±1 count. What is the error of measurement of the difference? Assume that random errors add.

1-12. You are a doctor in charge of a hospital testing lab. You send through a standardized sample of blood with 50 mg of iron per 100 ml of blood. The report comes back with 55 mg/ml. In the same run, a patient has a reading of 30 mg/ml. Report the patient's iron concentration by (a) correcting the results according to the standard (rounded to the nearest integer), (b) using the standard result as a bottom line error estimate.

2

ERROR ANALYSIS FOR ONE VARIABLE

In this chapter, we take a longer look at a set of data. We explore frequency distributions, different kinds of averages, and measures of errors.

2-1 FREQUENCY DISTRIBUTIONS

An instructive experiment is to show a box to a large class and get each person's estimate of the length of the box. This exercise brings home, in a graphic way, the uncertainty that is part of the measurement process. The results (in inches) of such an experiment, for a class of 100 persons, are as follows:

22 20 30 20 20 22 19 26 21 16 21 10 18 20 22 14 21 16 20 22 26 20 18
24 24 26 23 16 17 14 24 18 25 34 23 14 21 24 18 20 26 18 21 22 32 28
19 22 22 20 24 18 18 22 24 16 26 22 13 22 25 16 20 24 24 24 25 18 19
18 15 18 20 20 24 22 22 24 19 18 18 24 14 20 18 12 19 20 24 18 26 22
22 24 21 18 24 14 19 20

To make sense of this jumble of numbers, we make a tally:

10	/	23	//
11		24	///// ///// /////
12	/	25	///
13	/	26	///// /
14	/////	27	
15	/	28	/
16	/////	29	
17	/	30	/
18	///// ///// /////	31	
19	///// /	32	/
20	///// ///// ////	33	
21	///// /	34	/
22	///// ///// ////		

This tally also is called a *frequency distribution*, since it tells how frequently a particular value occurs. The tally resembles a graph of the data. Figure 2-1

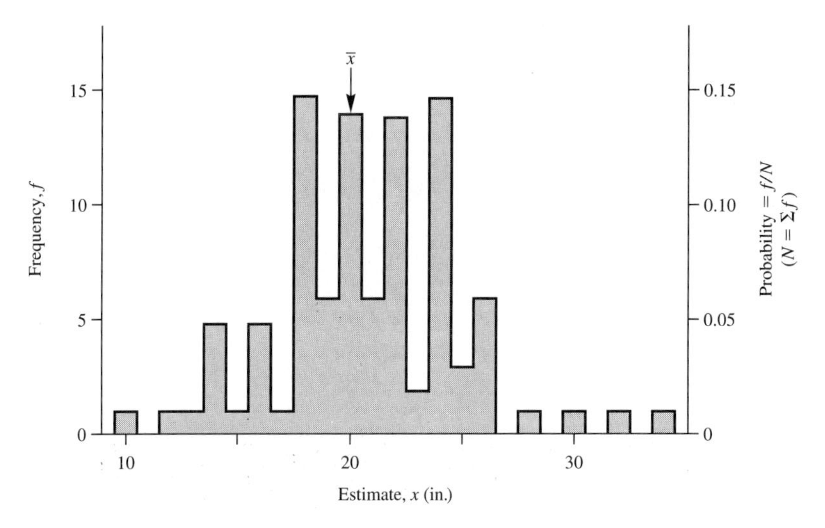

Figure 2-1 Frequency distribution: estimates of the length of a box by a class of $N = 100$ persons.

shows such a graph, with the values of length along the x-axis and the frequency of each value along the y-axis. Table 2-1 lists the values of this frequency distribution.

2-2 SOME PROPERTIES OF FREQUENCY DISTRIBUTIONS

We now consider some properties of frequency distributions.

Total number. The first property of a frequency distribution is N, the total number. This is simply the sum of the individual numbers of cases:

$$N = \Sigma f \tag{2-1}$$

By adding the f values in Table 2-1, we get the total number of persons in the class: $N = 100$.

The range. We define the *range* by the two extreme values. In the example, the range is 10 to 34 in. A closely related quantity is the *peak-to-peak deviation*, which we meet when looking at a noisy signal on an oscilloscope or other

TABLE 2-1 Frequency Distribution of Class Estimates.

Estimate: x	10	11	12	13	14	15	16	17	18	19	20	21	22	23	24	25	26	27	28
Cases: f	1	0	1	1	5	1	5	1	15	6	14	6	14	2	15	3	6	0	1

x	29	30	31	32	33	34	
f	0	1	0	1	0	1	

Mean = 20.76 in.
Standard deviation = 4.04 in.

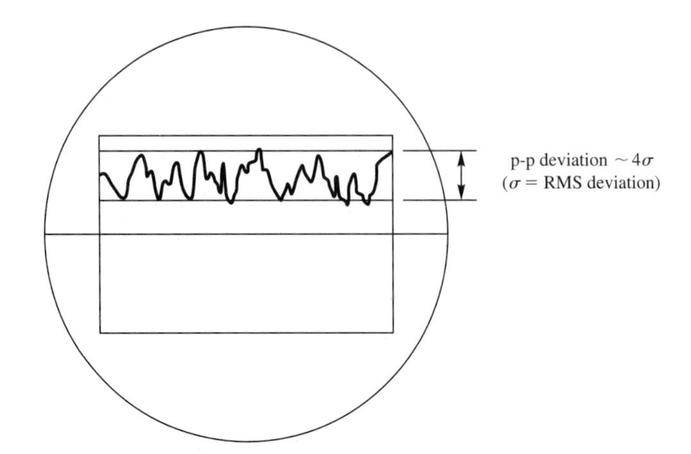

Figure 2-2 The RMS deviation of a noisy signal is approximately one-quarter of the peak-to-peak deviation.

signal-recording device, as shown in Fig. 2-2. The peak-to-peak (p-p) deviation is defined as the difference between the ends of the range of values on the screen or chart.

The average: mean, median, and mode. In Chapter 1, the mean was defined as the arithmetic average. Actually, there are three averages: the *mean*, the *median*, and the *mode*. We recall that the *mean* is the arithmetic average obtained by adding all the values and dividing by the number of cases N. A lengthy addition gives a value of 20.8 in. in the example. (To shorten the work, you can group the data, as shown in Section 2-6.) The *median* is the value that is higher than 50% of the data and lower than the other 50%. From the tally, we can see that the median is 20.5 in. in the example. The *mode* is the most frequently occurring value (the peak of the graph in Fig. 2-1). There is no clear-cut value of the mode in the example, as often happens.

The mean is simple to calculate with a small set of just a few pieces of data. For a large set of data, such as the example, the calculation is more work. In such a case, the median is easier to calculate and often will serve just as well for the average. The mode is less useful for actual data.*

The width of a distribution. Average and standard deviations. To say, "Our class guessed the length of the box to be about 20 $\frac{3}{4}$ inches," is not the whole story. It leaves out how people's estimates differed from each other.

*Social scientists find that many of the frequency distributions they meet are *skewed*, with much of the data piled at one end. Examples are personal income or wealth and life expectancy in underdeveloped countries. With this type of distribution, where it makes a big difference which average you use, the median is preferred. This type of problem seldom happens in physical sciences.

Likewise, a complete report of a laboratory measurement always gives the *error*, which is a measure of the "jitter" in the data.

The *range* is the quickest, but not the most useful, measure. In our experiment, the almost unbelievable values of 10 and 34 in., tell us about the extremes of the class, not the overall performance of most people.

The *average deviation*, which we met in Chapter 1, is easy to calculate. We simply take the average of the deviations from the mean; we take all values with positive signs. (If we kept the + and − signs, the result would be zero!) For example, the first five "measurements" in the box experiment (22, 20, 30, 20, 20 in.; mean = 22.4 in.) have deviations from the mean of −0.4, −2.4, +7.6, −2.4, and −2.4 in. The average deviation is (0.4 + 2.4 + 7.6 + 2.4 + 2.4)/5, which we shorten by grouping to be (0.4 + 3 × 2.4 + 7.6)/5 = 3 in., which we rounded to one significant figure. (With values ranging from 0.4 to 7.6, it is not meaningful to give the second digit after the decimal place. We will see in Section 2-3 that errors usually are only significant to one digit.)

A more commonly used measure of the spread, or width, of a set of data is the *standard deviation*, σ, also called the root-mean-square (RMS) deviation, which we get by adding the squares of the deviations, dividing by N, and then taking the square root:

$$\sigma_N = \sqrt{\frac{(x_1 - \overline{x})^2 + (x_2 - \overline{x})^2 + (x_3 - \overline{x})^2 + \cdots (x_N - \overline{x})^2}{N}}$$

We rewrite this in terms of the deviation $d = x - \overline{x}$:

$$\sigma_N = \sqrt{\frac{\Sigma(x - \overline{x})^2}{N}} = \sqrt{\frac{\Sigma d^2}{N}} \tag{2-2}$$

To test your calculator, try these three values of x: 1, 2, and 3. The standard deviation is 0.816. If you get 1.0, you have the *true*, also known as the *estimated*, or *unbiased*, or *sample* standard deviation σ_{N-1} given by

$$\sigma_{N-1} = \sqrt{\frac{\Sigma(x - \overline{x})^2}{N - 1}} = \sqrt{\frac{\Sigma d^2}{N - 1}} \tag{2-2$'$}$$

The observed standard deviation σ_N differs from the sample, estimated or true value because of a small effect. The observed mean $\overline{x}$ in expression (2-2) differs from the true mean. Expression (2-2′) corrects for this effect. Unless otherwise noted, we will always use the observed standard deviation σ_N in the text and simply write it as σ.

For example, we calculate the standard deviation of the first five data points in the class data:

$$\sigma = \sqrt{\frac{(22 - 22.4)^2 + (20 - 22.4)^2 + (30 - 22.4)^2 + (20 - 22.4)^2 + (20 - 22.4)^2}{5}}$$

which we abbreviate by grouping data:

$$\sigma = \sqrt{\frac{(22 - 22.4)^2 + 3 \times (20 - 22.4)^2 + (30 - 22.4)^2}{5}}$$

$$= 3.9 \quad \text{(rounded to 4 in.)}$$

The variance. We note here, for reference purposes, that the *variance* V is defined by the expression

$$V = \sigma^2 = \Sigma\frac{d^2}{N} = \Sigma\frac{(x - \bar{x})^2}{N} \tag{2-3}$$

Again, one has the population and sample variances, depending on whether N or $N-1$ is used in expression (2-3).

The frequency distribution of the mean. As discussed in Chapter 1, our common sense tells us that an average of several measurements is more *reliable* than a single meausrement. Imagine each person in the class giving you his or her length estimate on a slip of paper. Put all the slips in a hat and shake well. Now pull ten slips of paper from the hat, write down the numbers, calculate the mean, put the slips back, and shake again.

For example, the mean of the first ten values given in Section 2-1 is $(22+20+30+20+20+22+19+26+21+16)/10 = 21.6$. We notice that this average is closer to the mean of the total class (20.8 from Table 2-1) than most of the individual values. Next, we repeat this process over and over again, tally the results, and plot the frequency distribution of the means, as we did for the single entries (see Table 2-1 and Fig. 2-1). Table 2-2 and Fig. 2-3 show the results of 1000 drawings and means, done by a computer.

Note that the distribution of means is a bell-shaped curve. The mean of this new distribution agrees with the parent values (Table 2-1 and Fig. 2-1). However, the distribution of the means is narrower than that of the single pieces of data (compare Figs. 2-1 and 2-3).

Standard error of the mean. Statistical theory gives a result that agrees with computer "experiment"; the theoretically estimated *standard deviation of*

TABLE 2-2 Frequency Distribution of Mean ($N = 10$)

Mean: $\bar{x}$		13.5–13.9	14–14.4	14.5–14.9	15–15.4	15.5–15.9
Cases: f		1	1	1	0	0
x	16–16.4	16.5–16.9	17–17.4	17.5–17.9	18–18.4	18.5–18.9
f	4	5	9	12	42	40
x	19–19.4	19.5–19.9	20–20.4	20.5–20.9	21–21.4	21.5–21.9
f	90	100	147	162	118	113
x	22–22.4	22.5–22.9	23–23.4	23.5–23.9	24–24.4	24.5–24.9
f	73	44	22	16	0	0

Mean = 20.55 in; standard deviation = 1.44 in.

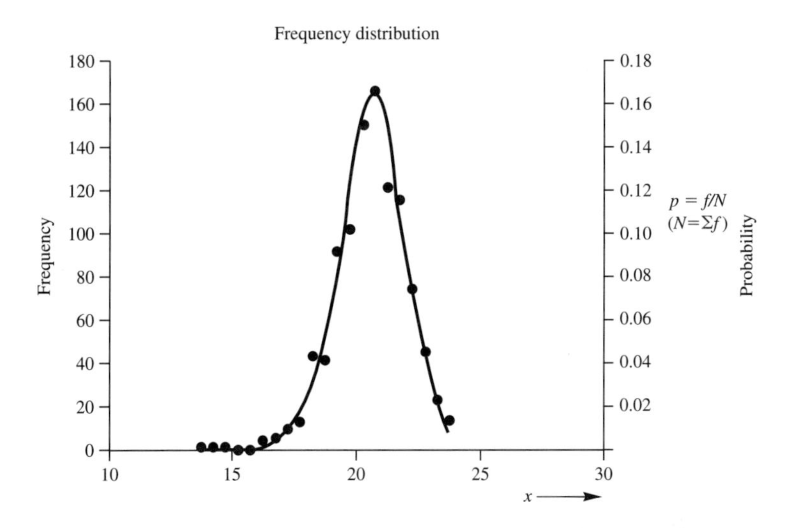

Figure 2-3 Frequency distribution of the mean of a sample of 10 from the distribution in Fig. 2-1. Total number of samples, $N = 1000$.

the mean σ_m is equal to the standard deviation of single observations, divided by the square root of $N - 1$, where N is the sample size:

$$\sigma_m = \frac{\sigma}{\sqrt{(N - 1)}} \tag{2-4}$$

The common term "standard error of the mean" means the same as standard deviation of the mean. In terms of the true standard deviation given by Eq. (2-2′), the standard error of the mean becomes

$$\sigma_m = \frac{\sigma_{N-1}}{\sqrt{N}} \tag{2-5}$$

Expression (2-5) is a law of diminishing returns. To halve the error, we must take *four* times as many observations. To cut the error by a factor of 10, we'd have to take 100 times as many measurements!

Example 2-1

A group of 10 students measures the length of a bar. Table 2-3 gives their results and shows the mean, standard deviation, variance, and standard error of the mean.

How reliable are estimates of the standard error? "Not very!" is the answer. Section 1-3 stated that *errors should be given only to one significant figure.* We will now see the reasons for that statement.

We have found in this chapter that taking measurements is like a public opinion pollster's sample. These measurements give us just an estimate of the true value we're trying to find. Likewise, the standard error is just an estimate. To

TABLE 2-3 Measurement of the Length of a Bar.

Measured Length (mm) x	Deviation from Mean $d = x - \overline{x}$	Square of Deviation d^2
100.0	−0.2	0.04
100.5	+0.3	0.09
100.3	+0.1	0.01
99.5	−0.7	0.49
98.6	−1.6	2.56
101.6	+1.4	1.96
99.2	−1.0	1.00
100.9	+0.7	0.49
101.1	+0.9	0.81
99.8	−0.4	0.16
$\Sigma x = 1001.5$ mm		$\Sigma d^2 = 7.61$ mm^2

Sample size: $N = 10$

Mean $\overline{x} = \dfrac{\Sigma x}{N} = 100.15$ mm, rounded to 100.2 mm

Variance: $V = \dfrac{\Sigma d^2}{N} = 0.76$ mm^2

Standard deviation: $\sigma = \sqrt{V} = 0.87$ mm, rounded to 0.9 mm

Standard error of

the mean: $\sigma_m = \dfrac{\sigma}{\sqrt{(N-1)}} = 0.29$ mm, rounded to 0.3 mm

Final result: $\overline{x} = 100.2(3)$ mm $= 100.2 \pm 0.3$ mm

illustrate this, we go back to our distribution in Section 2-1, with its standard deviation of 4.0 in. We had beginner's luck on the standard deviation we calculated from the first five persons: we got a value of 4 in.! But, if we were to take the next five persons, we would find a standard deviation of 3.3 in., which is not so lucky. We see that the measured standard deviation is as chancy, or is more so, as the mean itself. To get a better picture of things, we repeat our experiment of pulling ten slips of paper from the hat 1000 times. This time, we calculate the standard deviation of each set of data and let our tireless computer make a tally. Figure 2-4 shows a plot of the distribution of these standard deviations. The theoretical width of this distribution is given by

$$\sigma_\sigma = \frac{\sigma}{\sqrt{2(N-1)}} \qquad (2\text{-}6)$$

Even if $N = 10$, a rather large number of measurements, the theoretical expression (2-6) and Fig. 2-4 show that our estimate of the standard deviation, based on expression (2-2), is uncertain by about 25%. For a more typical student experiment, with $N = 3$ or 4, the error in σ is about 50%!

Thus, we draw the important conclusion: *error estimates are very rough. One significant figure is enough for the standard error.*

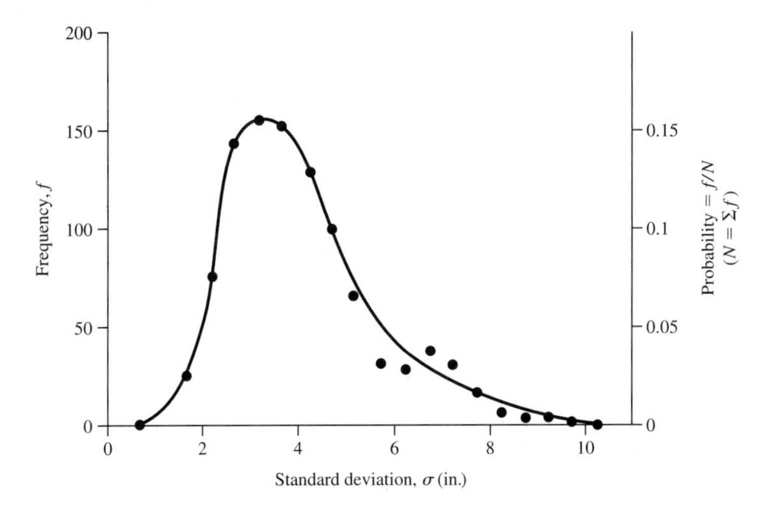

Figure 2-4 Frequency distribution of the standard deviation of a sample of 10 from the distribution in Fig. 2-1. Total number of samples, $N = 1000$.

In the next chapter, we will discuss error analysis for more than one variable, the study of how errors *propagate*, that is, how the effects of an error of measurement affect our calculations of other quantities based on this measurement. When you study this subject, take it all with a grain of salt. Elaborate error analyses are not justified by the basic crudity of the error estimate. Don't let yourself (or your laboratory instructor!) forget this. It will help you keep your perspective later!

2-3 PROPERTIES OF THE NORMAL DISTRIBUTION

The normal curve, also called the normal error curve, is the distribution approached by the mean as N becomes very large. This bell-shaped, symmetric curve has its mean, median, and mode at the center, at the origin of the x-coordinate. The y-coordinate gives the frequency for finding a mean at x. This curve is shown in Fig. 2-5.

Probability. The y-axes of Figs. 2-1 and 2-3 are relabeled "probability" (on the right side of the figure) by dividing the frequency by N, the total number of cases. The total probability, by this definition, is unity.

It is conventional to make the normal curve represent probability. The mathematical expression for the normal probability curve is

$$y = \frac{1}{\sigma\sqrt{2\pi}} e^{-(x^2/2\sigma^2)} \tag{2-7}$$

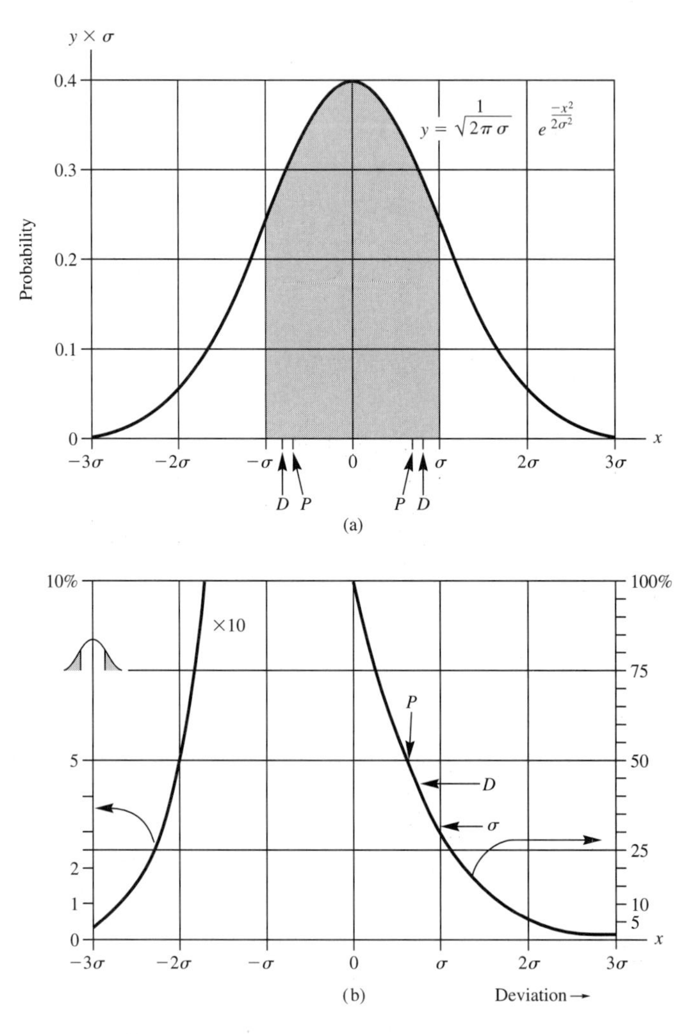

Figure 2-5 (a) The normal distribution curve. The shaded area is equal to the probability that a case deviates by one standard deviation σ or *less* from the mean: probability = 0.68. The points labeled P deviate by one probable error from the mean; the points labeled D are an average deviation from the mean. (b) Curves for the probability that a point deviates from the mean by *more* than the magnitude x (two-tailed test). The right curve is to scale; the left curve is magnified by a factor of 10.

The area A under the curve between two values of x is equal to the probability of finding an error in that interval. The entire area under the curve (from $x = -\infty$ to $x = +\infty$) is unity. Thus the area *outside* two values of x, the area in the tails (unshaded in Fig. 2.5), is $1 - A$. This is the probability that a measured value will lie outside the interval. Table 2-4 gives selected values of x and the probability that the error lies outside the values $\pm x$.

TABLE 2-4 Normal Probability Distribution

Deviation from Mean	Probability That Deviation is Exceeded	Comments
0	1	Value lies exactly at mean
$\pm 0.67\sigma$	$\frac{1}{2}$	Probable error
$\pm 0.80\sigma$	0.42	Average deviation
$\pm\sigma$	0.32	Standard (RMS) deviation
$\pm 2\sigma$	0.05	Significant difference
$\pm 2.6\sigma$	0.01	Highly significant difference

We will discuss each pair of points in the table. The first point, $x = 0$, lies *exactly* at the mean. According to the table, there is unit probability that this will *not* happen; there is no chance that a measurement will be exactly correct! The probable error, at $x = \pm 0.67\sigma$, gives you an even chance either way that it will or will not be exceeded. The chance that the error will lie outside the ± 1 standard deviation mark is about one-third. That is, about two-thirds of the measurements will lie inside the standard deviation marks; about one-third will fall outside.

The normal practice in quoting errors is to give the standard deviation of the mean, which gives you about 2 : 1 odds you'll be within your quote. Suppose you want to play safe and quote an error that is more unlikely to be wrong. A deviation that exceeds $\pm 2\sigma$ has one chance in 20 of happening. Such an event is called a *significant difference* by statisticians. Likewise, a 2.6σ deviation has only one chance in 100 of happening and is called *highly significant* by statisticians.

Recall that the normal distribution only holds for a very large number of measurements N. Most laboratory measurements involve small samples, with N as small as 2 or 3. In this case, the normal distribution is not valid and one should use the t distribution, which is given later in this chapter.

Peak-to-peak deviation versus RMS. A quick way to estimate the RMS (standard) deviation of a noisy signal is to use the approximate relation

$$\text{RMS deviation} \approx \frac{1}{4}(\text{peak-to-peak deviation}) \tag{2-8}$$

From Table 2-4, we see that expression (2-8) amounts to taking the peaks at the 95% probability values. This approximation works surprisingly well.

2-4 THE BINOMIAL DISTRIBUTION

Sampling. A common statistical problem arises from sampling. One example is the jelly bean production line manager who takes a random sample of 50 to check for the distribution of colors. Each sample is different. Is the machine that puts in each color broken, or is the variation due to the randomness of the sample?

Public opinion polling, such as election prediction, is another example. The pollster might walk down a New Haven street and ask the first 100 voters he or she meets, "Whom do you prefer for president in the next election?" If 60% of the persons who have made up their minds choose the Democratic candidate, does that mean the Democrat will win? Since New Haveners have voted for the Democratic slate without exception for many years, the answer is clearly no: the sample is systematically unrepresentative of the U.S. population. Suppose a representative sample of persons (chosen by race, income, state, age, etc. to eliminate systematic sampling errors) still is 60% Democratic. What are the chances the Republican will win? One might guess that, with a sample as small as 100, the chances are pretty good. (See problem 13.)

The binomial distribution predicts the outcome of sampling measurements, such as those of the jelly beans. Suppose that the jelly bean making machine accurately counts out 30 brown pieces out of every 100. However, these get mixed in packing, so that we cannot expect the first 100 jelly beans we take to have exactly 30 browns. We say that, for a random sample of a portion of the mixture, the probability p of a jelly bean being brown is 0.3 (30%) and the probability of it being not brown (or any other color) is $q = 1 - p = 0.7$ (70%). The total probability of being brown or not brown is $q + p = 1$. The binomial theorem, proven in algebra texts, says that, in a sample size of N, the probability P of getting s brown jelly beans and $N - s$ jelly beans that are not brown is given by

$$P = {}_N C_s p^s q^{N-s} \tag{2-9}$$

where

$${}_N C_s = \frac{N!}{s!(N - s)!}$$

is the number of combinations of N things taken s at a time and $N! = N(N-1) \cdots (2)(1)$ (N factorial).

Almost all calculators have buttons for factorials and powers of quantities. Most have ${}_N C_S$. Thus, most calculators make it simple to calculate the binomial distribution.

Example.

Calculate the probability of getting s brown jelly beans ($p = 0.3$) in a random sample of 10. The data are shown in Table 2-5 and also in a graph (Fig. 2-6).

Connection among the distributions and other interesting relations. It can be shown that the standard deviation σ of the binomial distribution is $\sqrt{Npq}$. This has several consequences:

TABLE 2-5 Binomial Probability Distribution: $N = 10, p = 0.3$.

s	$N - s$	${}_NC_s$	$P = {}_NC_s p^s q^{N-s}$
0	10	1	0.028
1	9	10	0.121
2	8	45	0.233
3	7	120	0.267
4	6	210	0.200
5	5	252	0.103
6	4	210	0.037
7	3	120	0.009
8	2	45	0.001
9	1	10	0.000
10	0	1	0.000

1. The error of a single measurement, equal to σ, is proportional to the square root of the sample size. Thus, if instead of 50 pieces you chose a sample of 200 jelly beans, the error of your measurement of the number of browns would double in size.
2. The error in the *proportion* (fraction $= p$) of the population is equal to

$$\sigma_p = \frac{\sigma}{N} = \sqrt{\frac{pq}{N}} \tag{2-10}$$

inversely proportional to the square root of N. Thus, in your larger sample of 200, your estimate of the *proportion* of red jelly beans would have an error only half of that for a sample of 50. This is an example of the general rule that statistical errors of averages (mean, proportion, etc.) are inversely proportional to the square root of the sample size or of the

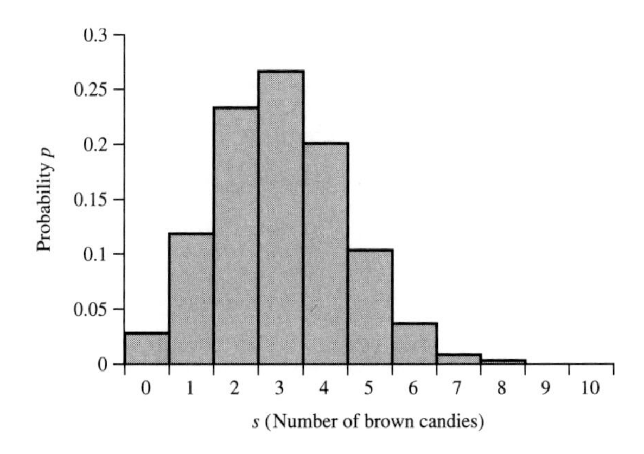

Figure 2-6 Probability of s events in $N = 10$ trials: $p = 0.3$.

number of measurements, which we already have seen for the standard error of the mean.

3. For a large sample size ($N \gg 1$) and for a small probability for p ($p \ll 1$, $q \approx 1$), the binomial distribution approaches a limit called the Poisson distribution, which is discussed in Chapter 5. The standard deviation becomes just the square root of the number counted $\sigma = \sqrt{pN}$.
4. In the limit that Np is a large number ($Np \gg 1$), the binomial distribution approaches the normal distribution.

2-5 THE *t* DISTRIBUTION

Imagine that you have measured the acceleration of gravity in three trials to be 9.5, 9.6 and 9.7 m/sec^2, with a mean value of 9.6 m/sec^2 and a standard error of the mean $\sigma_m = 0.06$ m/sec^2. The accepted value is 9.8 m/sec^2. What do you say in your lab report? Is the difference of 3 standard deviations (SDS) between your result and the accepted value real or is it a statistical fluke?

Table 2-4, based on the normal distribution, lists 3 SD as "highly significant," likely to happen in less than 1% of the time purely by chance. However, it is wrong to use the normal distribution, which is only correct for a large sample size; you only measured *g* three times.

"Student's" (pen name for William S. Gossett, a statistical pioneer) distribution handles the problem of estimation of significance for small sample sizes. The "*t* ratio" of the difference/SD has two sources of uncertainty: the difference itself and the SD, which also is chancy (see Section 2-2). The "*t*-test" replaces Table 2-4 with a separate column for each sample size.

Table 2-6 shows that a difference of 3 SD for $N = 3$ data points is not significant. Thus this deviation could have happened through chance ($p > 0.05$).

TABLE 2-6 The t-Test for Significance.*

	Deviation (in Standard Deviations)				
Number of Points	2	3	4	5	∞
Probability that deviation is exceeded					
1	0.0	0.0	0.0	0.0	0.0
$\frac{1}{2}$	1.0	0.8	0.8	0.7	0.7
$\frac{1}{3}$	1.7	1.3	1.2	1.1	1.0
0.05 (significant)	12.7	4.3	3.2	2.8	2.0
0.01 (highly significant)	63.7	9.9	5.8	4.6	2.6

* For additional values, use the functions TINV and TDIST in most spreadsheets. TTEST tests whether two sets of measurements have different means.

Table 2-6 also shows that it would take a 10 SD difference to be highly significant for a sample size of three measurements!

2-6 COMPUTATION: MEAN, STANDARD DEVIATION, AND STANDARD ERROR OF THE MEAN

This section treats the practical details of computing the quantities that are useful in processing laboratory data. Programs for these calculations are given in Appendix B.

The Round-off Crash

A problem could arise in the rare situation when you have a set of large numbers that differ only slightly from each other. The reason is round-off error. (See the discussion in the Preface.) To test this effect, put these three numbers in your calculator and see what comes out: 1,000,001, 1,000,002, and 1,000,003. From the example on page 13 (1, 2, and 3), you know the correct answer is 1,000,002 for the mean and 0.816 for the standard deviation (1.0 for estimated standard deviation). If your calculator gives you these answers, you're lucky and have no worries. If it gives you the wrong answer for the standard deviation (usually zero), you must avoid the round-off crash by subtracting the large part of your numbers (1,000,000 in this example), calculating the mean and standard deviation, and then correcting your mean by adding back the big number. (The standard deviation is unaffected by this subtraction.)

Mean, standard deviation and error of mean: single entry. With any scientific calculator, you have no program to learn: the program is "hard-wired" into the calculator. Nevertheless, you must remember these points:

1. Always clear the statistical registers (N, $\Sigma\, x$, and $\Sigma\, x^2$) before you put in your new data. If you don't, you'll learn the meaning of the acronym GIGO ("garbage in, garbage out").
2. Use a simple sample set of data ($1, 2, 3, \sigma = 0.816$) to see which standard deviation you get. If the answer is 1.0, see Section 2-2 for information on the estimated standard deviation σ_{N-1}.
3. If you are in the unusual situation that you have a set of large numbers that are close together, reread the discussion in the paragraph on the round-off crash.

Handling large sets of data by tallying or grouping. For larger numbers of data points, the work is tiresome and likely to lead to mistakes. Almost all scientific calculators will save labor and avoid mistakes by taking data that are sorted into groups. This section gives shortcuts that cause very little loss of accuracy in data processing.

Tally method. Use a tally to arrange the data in numerical order, as in Table 2-1. Use the median, the value that divides the data in halves, as your estimate for the mean. Find the two points, equally spaced from the mean, that include about two-thirds (more precisely, 68%) of the data. The standard deviation is half the separation of these points. There is no need to group the data with this method; if the data happen to be grouped (as in Table 2-1), assume the data points are equally spaced within the interval.

Example 2-2

Find the mean and standard deviation of the tallied data in Table 2-1. A graph of the data, with the grouping shown, is given in Fig. 2-7.

For convenience, we use a piece of graph paper, use one division for each data point, and use brackets to mark the intervals of x (see Fig. 2-7). We will now see how to find the points that correspond to $\bar{x}, \bar{x}+\sigma, \bar{x}-\sigma$. Since the class rounded off lengths to the nearest integer, we can consider the data grouped in intervals 9.5–10.5, 10.5–11.5, 11.5–12.5, 12.5–13.5, 13.5–14.5, and so on.

The median is midway between the 50th and 51st points. By simple counting, we find that it is halfway between 20 and 21. Therefore, we set $\bar{x}=20.5$.

For the standard deviation, we take half the separation between the 16th and 84th points (32% of the 100 data points lie outside). We find the 16th point near the left of the group of 15 points between 17.5 and 18.5. We give it the value of 17.5. Likewise, since the 84th point is near the top of the 15 points lying between 23.5 and 24.5, we give it the value of 24.5. The standard deviation σ is half the separation of these two points: $\sigma=(24.5-17.5)/2=3.5$.

These two values compare fairly well with the exact values of $\bar{x}=20.76$ and $\sigma=4.04$, which we round to 4.

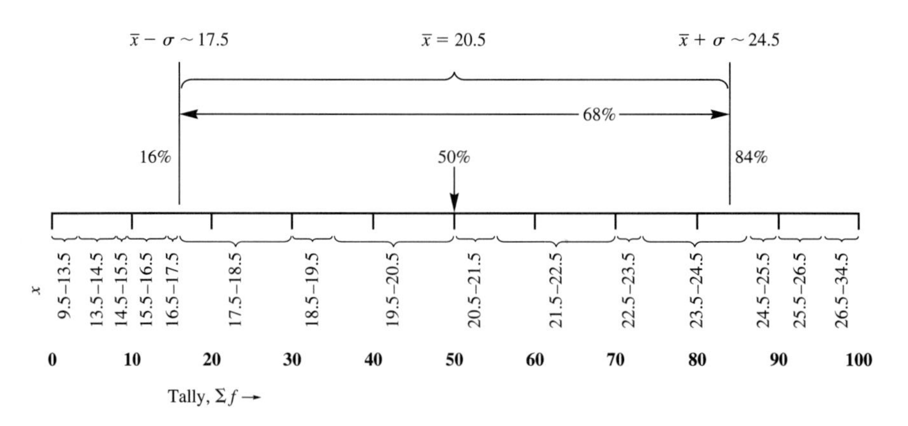

Figure 2-7 The tally method of finding the mean and standard deviation of a large set of data.

Statistical calculations by grouping. In this method, all f pieces of data that fall within a certain interval are lumped together and are given the value of x that is in the center of the interval. This procedure usually is quite accurate, as long as the intervals are less than one standard deviation wide. Breaking the entire set of data into about eight to ten equally spaced sets usually is enough.

Each value of x that goes with the center of the interval is multiplied by f in all calculations. The total number of data points N simply is the sum of the fs, which sometimes are called the *weights*:

$$N = \Sigma f \tag{2-11}$$

$$\overline{x} = \Sigma \frac{fx}{N} \tag{2-12}$$

$$\sigma = \sqrt{\Sigma \frac{f(x - \overline{x})^2}{N}} \tag{2-13}$$

Example 2-3

Group the data in Table 2-1 to find the statistical properties.

We divide the data into groups, each 3 in. wide (Fig. 2-8). Table 2-7 shows the computation of the statistical properties by means of expressions (2-11), (2-12), and (2-13).

TABLE 2-7 Grouping and Calculation of Statistical Properties of Class Estimates of the Length of a Box. Data from Table 2-1.

Length					
Range	x	Weight f	Weighted Length, fx	Deviation $(x - \overline{x})$	Weighted Square $f(x - \overline{x})^2$
9.5–12.5	11	2	22	−9.69	187.8
12.5–15.5	14	7	98	−6.69	313.3
15.5–18.5	17	21	357	−3.69	285.9
18.5–20.5	20	26	520	−0.69	12.4
21.5–24.5	23	31	713	+2.31	165.4
24.5–27.5	26	9	234	5.31	253.8
27.5–30.5	29	2	58	8.31	138.1
30.5–33.5	32	1	32	11.31	127.9
33.5–36.5	35	1	35	14.31	204.8

$N = \Sigma f = 100 \qquad \Sigma fx = 2069 \qquad \Sigma f(x - \overline{x})^2 = 1689.4$

$\overline{x} = \dfrac{\Sigma fx}{N} = 20.69$, rounded to 20.7; $V = \dfrac{\Sigma f(x - \overline{x})^2}{N} = 16.894$

$\sigma = \sqrt{V} = 4.11$, rounded to 4

$\sigma_m = \dfrac{\sigma}{\sqrt{(N - 1)}} = 0.413$, rounded to 0.4

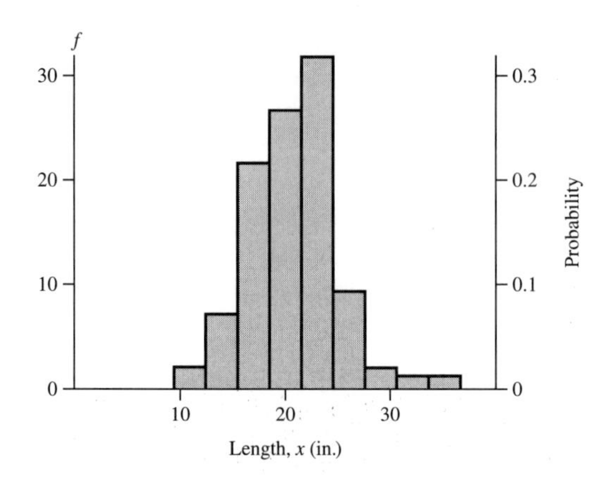

Figure 2-8 Bar graph of the data from Fig. 2-1, grouped in intervals of 3 in.

Sample data for grouped calculations. Test your program/calculator on this set of sample data:

x:	1	2	3	4	5	$N = 68$	$\Sigma fx^2 = 204$	$\Sigma fx = 680$	Mean $= 3$	$\sigma = 1.0$
f:	4	18	24	18	4	σ_{N-1}	$=$ 1.0074	Standard error of mean $= 0.122$		

Summary

The standard deviation of a set of data is defined as

$$\sigma = \sqrt{\frac{\Sigma\, d^2}{N}}$$

The estimated standard error of the mean is

$$\sigma_m = \frac{\sigma}{\sqrt{(N-1)}}$$

In the normal distribution, about two-thirds of the measurements will lie within the errors. Errors should be quoted to one significant figure. Data can be grouped to calculate the

$$\text{mean } \bar{x} = \frac{\Sigma\, fx}{N}$$

and

$$\text{standard deviation } \sigma = \sqrt{\frac{\Sigma\, fd^2}{N}}$$

The binomial distribution gives the chances of observing a given number of cases, with a N trials and a probability p. The standard deviation of the binomial distribution is $\sqrt{Npq}$. The standard deviation of the probability p is

$$\sigma_p = \frac{\sigma}{N} = \sqrt{\frac{pq}{N}}$$

A difference of experimental and accepted value should be at least twice the error to be considered significant. In case of small samples, the t-test shows that the ratio should be even higher.

Problems

2-1. A set of measurements for the time it took a body to fall 1 m was 0.45, 0.42, 0.41, 0.48, and 0.44 sec. Find each of the desired quantities, rounded off correctly: (a) number of observations n, (b) average (mean), (c) median value, (d) range of values, (e) standard deviation of the data, (f) standard error of the mean, (g) fractional error of the mean, (h) percentage error of the mean.

2-2. A spaceship counted the number of dust particles detected per minute in a cometary atmosphere. Table 2-8 lists the frequency of each number of counts. Use the tally method (Section 2-6) to find (a) the median number of counts, and (b) the standard deviation of the distribution. By actual calculation, find (c) the mean of the distribution, and (d) the standard deviation of the distribution. Compare your answers to parts (a) and (c) and parts (b) and (d).

TABLE 2-8 Count Frequency Distribution of Cometary Dust Particles.

Number of counts	0	1	2	3	4	5	6	7	8	9	10	11	12 or more
Frequency	1	3	9	15	15	17	17	10	8	8	4	2	0

2-3. As a test of statistical theory, take the first 25 pieces of data from the distribution at the beginning of this chapter. Find the (a) mean and (b) standard deviation of the data. Then take the data, break them up into five groups of five data points each, and calculate (c) the mean of each group, and (d) the standard deviation of this new set of data. Then calculate (e) the expected standard error of the mean from part (b) and expression (2-4). Compare your results from parts (d) and (e).

2-4. From Fig. 2-5(b), give the probability that the magnitude of the observed deviation from the mean is (a) less than one standard deviation, (b) less than half of a standard deviation, (c) exactly equal to zero, and (d) greater than 2.5 standard deviations. *(e) Give the probability that the deviation is positive and greater than three standard deviations.

2-5. Probabilities are sometimes stated as odds. If the odds are 2 to 1 in favor of an event happening, it has a probability of $\frac{2}{3}$. If the odds are 3 to 1 against it, it has a probability of $\frac{1}{4}$. What are the odds that a given event will differ from the mean (a) in magnitude by more than one standard deviation, (b) in magnitude by the probable error or less, *(c) by +1 standard deviation or more, and (d) by more than 2.6 standard deviations in magnitude?

2-6. Which is the better way to measure the period of a pendulum, (1) to time one period ten times and take the average, or (2) to take a single measurement of ten periods? To find the answer, solve this problem: A single period is 2 sec, with a random standard error of measure of $\sigma = 0.1$ sec. Calculate (a) the standard error of the mean of ten measurements, (b) the percentage error, and (c) the percentage error in one measurement of ten periods (20 sec).

Next assume that there is a systematic error of +0.05 sec; that is, because of reaction time, parallax, or other factors, the time measured is on the average 0.05 sec too long. Find the percentage error due to that cause for (d) ten measurements of a single period (2 sec each), and (e) a single measurement of ten periods (20 sec).

2-7. A single measurement of a quantity has an error σ. How many measurements must you take so that the error of the mean is (a) $\sigma/2$, and (b) $\sigma/10$?

***2-8.** Show that if you make two measurements of a quantity, the standard error of the mean is half the difference between your two values.

***2-9.** Show that the relation between the true or estimated standard deviation (sometimes called the population standard deviation) and the observed σ_N is $\sigma_{N-1} = \sigma_N\sqrt{N/(N-1)}$.

***2-10.** Check the rule of thumb given in Chapter 1, that the average deviation of a set of three measurements is a good approximation to the error (standard error of the mean). Given a true distribution standard deviation σ, find (a) the standard deviation of the mean for a data set of $N = 3$. Use Table 2-4 and expression (2-2′) (or the result of Problem 2-9) to find (b) the average deviation for the same data set; remember that you must use σ_N for the observed deviations. Compare your answers to parts (a) and (b).

***2-11.** Use the grouped data in Table 2-1 to calculate the (a) average deviation of the data. Use the tally at the beginning of the chapter to estimate (b) the probable error for the same distribution (half the data lie within the two points mean ± one probable error). Compare your results with the (c) average deviation and (d) probable error calculated from the standard deviation given in Table 2-1 and the contents of Table 2-4.

2-12. You have just made two measurements of a quantity and you found the mean, the standard error of the mean, and the deviation of your mean from the accepted value. What should be the t ratio (deviation/SEM) for the deviation to be a significant difference?

2-13. If the sample on page 20 were representative of the U.S. population, what would be the chances (probability) of a Republican victory?

3

ERROR ANALYSIS FOR MORE THAN ONE VARIABLE

In this chapter, we arrive at one of the trickiest parts of data processing: error analysis. Error analysis is concerned with two related topics: first, how to combine more than one measurement. An example of this is to measure the length L and width W of a rectangle to get its area $A = L \times W$. The second is the *propagation of errors*, in which you measure a quantity x, go through a calculation involving x to get a result y, then scratch your head and wonder, "What is the effect of my error in x on my value of y?" An example of this would be measuring the side of a square x, calculating its area $A = x^2$, and then asking, "What is my error in the area?" This chapter discusses these topics and gives simple methods (in Section 3-5) of avoiding the complicated computations of error analysis.

3-1 HOW ERRORS ADD

Independent versus correlated errors. To understand how errors add (or to speak more correctly, how errors *combine*), do an imaginary experiment. Suppose you have a cubic hunk of metal, about 1 in. on a side, and a micrometer to measure it. You measure the distance between opposite sides and get 1.001 in.

Suppose your "hunk of metal" is in reality a superaccurate 1-in. gage block, accurate to a few millionths of an inch. This tells you your error of measurement was 0.001 in. or, in more handy percentages, $100 \times (0.001/1.000) = 0.1\%$.

Next, suppose you want to know the area of one face. You get out your calculator, multiply 1.001 in. $\times$ 1.001 in. = $(1.001 \text{ in.})^2$ = 1.002001 in.^{2}, which you round to three decimals to get 1.002 in.^{2}. What about the error in area? Here's a surprise. The percentage error is $100 \times (0.002 \text{ in.}^2/1.000 \text{ in.}^2) = 0.2\%$, double that in your original measurement! How did that extra tenth of a percentage creep in to double your error?

To answer that question, we recall that a square is also a rectangle, with area $A = L \times W$. We *assumed* both length and width to be equal: $L = W = 1.001$, multiplied them together, and got the answer, as shown in the arithmetic in the last paragraph. By doing this, we forced *both* the length L and width W to have the same error of 0.1%; the error then appeared *twice* in the calculation. Thus the mystery of the doubled error is solved.

This is an example of *correlated* or *nonindependent errors*. The errors in L and W were exactly the same. We made them so by our method of calculation. Whenever a single measurement enters more than once into two or more different quantities, we have to take this effect into account.

Independent measurements. Addition of random errors. Suppose we had measured length and width of the cube separately, to an error of 0.1%. Next we ask, what would be the error in the area? We consider all possibilities. The first guess is the pessimistic one: we might imagine that the errors simply would add; the net result is a doubling, which leaves us just where we were before. The second guess is the optimistic one: the "law of averages" says that for every positive error there is likely to be a negative error. Thus the second measurement, on the average, will *cancel* the first one, with a net error of zero! A third possibility could be "none of the above," somewhere in between wild optimism and pessimism.

We phrase the question more sharply: How do two or more totally independent, random measurements combine? By now, you know how we'll find the answer. We'll do a computer experiment. The name of that experiment will be "the drunkard's walk."

The drunkard's walk. Imagine a *random walk,* the path of a staggering drunk. He takes a step one way, sways back and forth, takes a step in another direction and so forth. We imagine the walk to be truly random: the direction of a given step is totally unrelated to that of the preceding one, as is shown in Fig. 3-1.

Figure 3-2 shows the first few steps of such a walk, calculated by a computer, which also made the direction of each step random and, incidentally, randomized the step length. For a demonstration of a random walk with steps of constant length, run the cT program *adderr* (on the enclosed disk).

We notice that the pessimistic guess is certainly wrong. If the drunkard took N steps, by this guess, he would be a distance N times the length of one step from his starting place. This is obviously not the case; he isn't getting anywhere that fast! The only way the "errors" could add to a distance of N steps would be for him to sober up and walk a straight line.

Random Errors Do Not Simply Add Arithmetically

The figure gives some hope to the optimist (guess number 2); after 14 steps, the drunk comes back almost exactly to his starting place. But this is only dumb luck. Figure 3-3 shows the distance he travels after thousands of steps

Figure 3-1 A "random walk" is a walk in which a person, instead of following a straight line, continuously changes his direction by turning at unpredictable angles. From George Gamow, *Matter, Earth and Sky*, 2nd ed., © 1965, p. 217. Reprinted by permission of Prentice-Hall Inc, Englewood Cliffs, N.J.

(assuming he is still drunk and conscious!). Falteringly, but inexorably, the distance increases. The result is that the distance grows neither as N (the pessimist's guess) nor hovers around zero (the optimist's), but is "none of the above"! In fact, Fig. 3-3 shows that the distance *increases as the square root* of the number of steps (i.e., as $\sqrt{N}$). This result is important in that it governs more than the drunkard's walk: it underlies the theory of errors, the

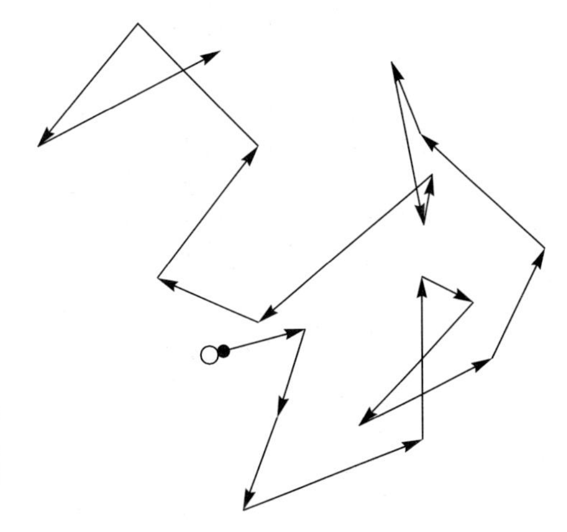

Figure 3-2 Computer simulation of a random walk, which starts at the origin 0. The x- and y-coordinates of each step have random values between ± 1.

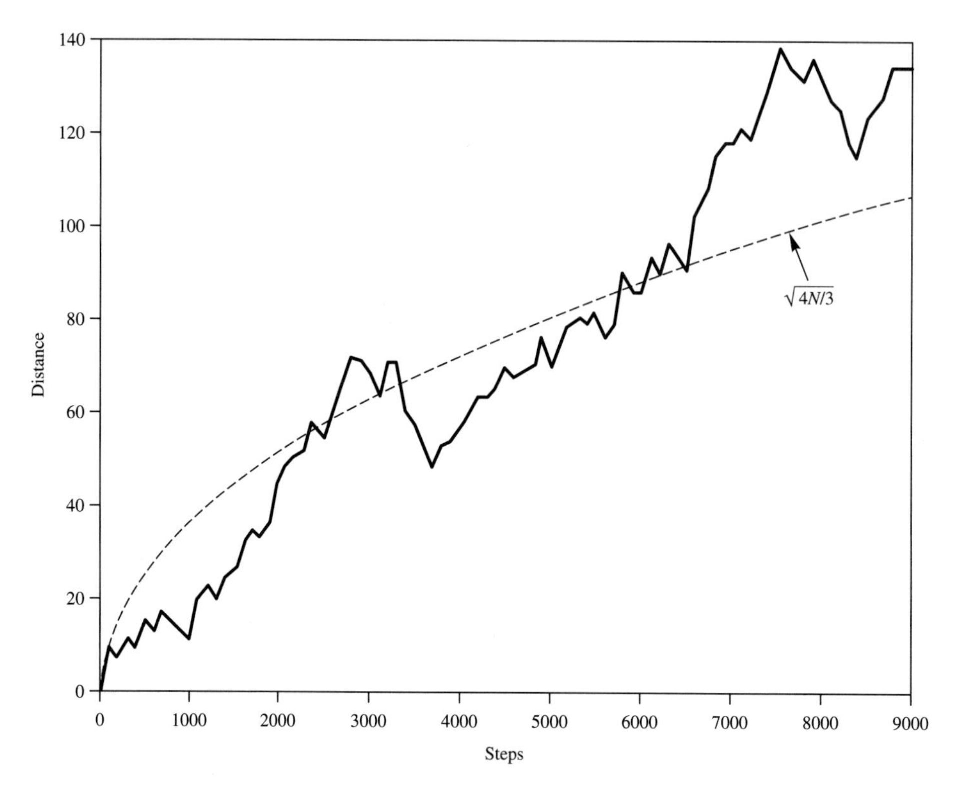

Figure 3-3 Distance from the origin of the random walk in Fig. 3-2 after N steps. The broken line is the theoretical value, which is a statistical average.

diffusion of molecules in a gas or liquid, the growth of explosions or other catastrophic breakdowns, and many other natural processes.

We now return to our original question: How do random errors add? To find an answer, we examine the square root of N rule for the simplest possible case: two equal, random errors. We call them A and B, where $A = B$, and now ask a more pointed question: How can two equal quantities add, such that their sum can be $\sqrt{2}$ times each? In other words, How can $A + A = \sqrt{2}A$? The answer is shown in Fig. 3-4. If we combine the errors, as the legs of an equilateral, right triangle combine to form the hypotenuse by the Pythagorean theorem, we get our answer. Two random errors A and B combine (on the average) by the root-mean-square sum to form the resultant error C:

$$C = \sqrt{A^2 + B^2} \tag{3-1}$$

Expression (3-1) holds for the error in the *sum* of two quantities. What about the error in the *difference* in two quantities? How do we *subtract* two random errors? The answer is, "It's just the same as *adding* two random errors;

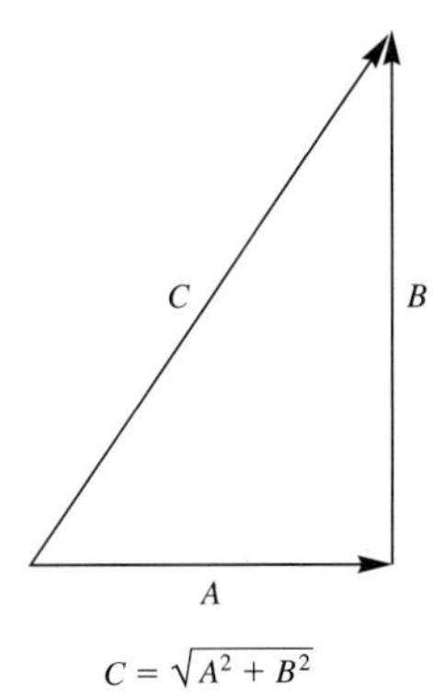

Figure 3-4 Two random, independent errors A and B add according to the Pythagorean theorem (i.e., like perpendicular vectors). On the average, the sum $C = \sqrt{(A^2 + B^2)}$.

the word random means you can't distinguish between positive and negative deviations." Thus, expression (3-1) holds, whether we want the error in the sum or difference of two quantities.

Example 3-1 Measurement of the Length of an Object

Compare the relative precision of two methods of measuring the length of an object:

1. By lining up one end of the ruler with an end of the object (assume this process can be done exactly)
2. By deliberately *not* lining up the ends, but by measuring the difference between the readings for each end of the object.

Solution. The random error of measurement is *increased* by 40% by using the second method instead of the first. To see this, imagine that the measurement process has an error of 1 mm, the object is exactly 80 cm long, and 5 mm are worn off one end of the ruler. Then, by method 1, the measurement will be 80 cm + 0.5 cm (systematic error because of the worn end) ± 0.1 cm (random error of measurement). The result is 80.5 ± 0.1 cm.

Next, consider method 2. Let one end be at 10.0 ± 0.1 cm; let the other be at 90.0 ± 0.1 cm. The difference is 80.0 cm. The error in the difference of the two quantities, by the discussion of the last paragraph, is given by expression (3-1): the error in the difference is $\sqrt{(1^2 + 1^2)} = 1.4$ mm. The random error of the second measurement is larger by 0.4 mm, although the systematic error of +5 mm is not present.

3-2 PROPAGATION OF ERRORS. SINGLE MEASUREMENT

Relative (fractional) and percentage errors. In this section, we assume a measurement of only one quantity x and ask how the standard error σ_x in x propagates in different functions of x. It is convenient to consider not only the error itself, but also the *relative (fractional) error* σ_x/x and *percentage error*: % error in $x = 100\sigma_x/x$. In error theory, we always consider the fractional error to be small compared to 1: $\sigma_x/x \ll 1$. (Large fractional errors are very unusual in physics laboratories.) All the expressions that follow are based on that assumption.

Propagation of errors: results. That section lists the results for propagation of errors for a function $z(x)$, where x is a single, measured variable.

1. Product of a constant and variable. Suppose you have a tiny postal scale that only weighs to 1 ounce and you want to mail a ream (500 sheets) of paper. How much does it weigh? You cleverly take a packet of five sheets of paper, put them together, and find the total weight to be 1 oz. Under the assumption that all sheets of paper have the same weight, the weight of a ream is 100 oz, or 6 lb and 4 oz. You put on enough postage for a 7-lb package.

Will your package make it? That depends on the error of your estimate of the weight. Suppose your weighing had an error of 0.1 oz. Then the ream would have an error of 100×0.1 oz $= 10$ oz. At most, your package would weight 6 lb 14 oz, and you are safe (if the wrapping doesn't weigh too much).

In this example you multiplied your measured quantity and its error by 100. You can readily imagine the reverse, in which you have only a large set of scales, weigh a ream of paper in pounds, and find the weight of a five-sheet letter by division. In this case, you would divide both your result and error by 100.

To summarize, *when a measured quantity is multiplied (divided) by a constant, the absolute error is likewise multiplied (divided) by the same constant.* To put it in mathematical language, if $z = ax$,

$$\sigma_z = a\sigma_x \tag{3-2}$$

Percentage Error. Note that, in the example, the *fractional* error was 0.1 oz/1 oz = 10 oz/100 oz = 0.1, which is the same for both quantities. Likewise, the *percentage* error is 10% in both cases.

We further conclude that *multiplying (dividing) a measured quantity by a constant leaves the fractional (percentage) error unchanged.* Again, in mathematical language, if $z = ax$,

$$\sigma_z/z = \sigma_x/x \tag{3-3}$$

Likewise,

$$\%\text{ error in } z = \%\text{ error in } x \tag{3-4}$$

2. Variable to a power. We saw, in the example at the beginning of the chapter, that squaring a quantity doubled its *fractional (percentage)* error; by a similar calculation, you will find that cubing it triples the *percentage* or *fractional* error. By means of calculus or the binomial theorem, we can prove that *raising a quantity to the nth power multiplies its percentage (fractional) error by n.* In mathematical language, if $z = ax^n$, the relative error in z is given by

$$\frac{\sigma_z}{z} = \frac{n\sigma_x}{x} \tag{3-5}$$

and the percentage error by a similar expression:

$$\% \text{ error in } z = n(\% \text{ error in } x) \tag{3-6}$$

3. General case. If z is any function of x, $z = z(x)$, the error in z is given by calculus:

$$\sigma_z = \frac{dz}{dx}\sigma_x \tag{3-7}$$

Derivations of these expressions can be found in the references.

Finding Errors with a Calculator.

If you find formulas (3-2) to (3-7) tiresome, you can use your calculator to get around the mathematics, as we did in Section (3-1) for finding the error in the area of a square. We generalize this to any function z of a measured variable x:

Given: a measured value x with its error σ_x,

To find: z and σ_z, where z is any function of x.

Solution: Use your calculator to find z from x. Then use the same calculation to find $z \pm \sigma_z$ from $x \pm \sigma_x$, from which you can find σ_z by subtraction. (In most cases, only two calculations suffice: either x and $x + \sigma_x$ or x and $x - \sigma_x$.)

Example 3-2.

The period T of a pendulum is given by the formula $T = 2\pi\sqrt{\frac{L}{g}}$, where L is the length in meters and $g = 9.8$ m/sec^2. If $L = 0.500 \pm 0.005$ meters, find the period and its error.

Solution. We first find a convenient formula: $T = 2.007\sqrt{L}$. We then put the values L, $L + \sigma_L$, $L - \sigma_L$: 0.5, 0.505, and 0.495 into the formula. The results are $T = 1.4192$ sec, 1.4263 sec, and 1.4120 sec, respectively. The differences are 1.4192 sec − 1.4263 sec = 0.0071 sec and 1.4263 sec − 1.4120 sec = 0.0072 sec, both of which round to 0.007 sec. The final result is $T = 1.419 \pm 0.007$ sec.

We note that either error calculation would have been enough.

3-3 PROPAGATION OF ERRORS. MORE THAN ONE MEASUREMENT

1. Addition or subtraction. We now come to the problem of combining two independent measurements. We already saw in Section 3-1 that, if the quantities are added or subtracted, the errors obey the Pythagorean theorem [expression (3-1) and Fig. 3-4]. We can state this result mathematically: if $z = x \pm y$, then

$$\sigma_z = \sqrt{(\sigma_x^2 + \sigma_y^2)} \tag{3-8}$$

2. Multiplication or division. The shrewd reader might object to the way we found the area of a square gage block in Section 3-1 by squaring the length of one side. "How do you know both sides are equal? Shouldn't you have measured both the x and y sides and then *multiplied* x and y to find the area $A = xy$?" The answer to that question is, "Of course, you are right. Now we'll see how to combine the errors in x and y when we multiply them together."

To do this, we find the errors due to x and y separately by holding the other constant. We then add the errors in the usual fashion.

First, we imagine y held constant. Then $A = cx$, where c is a constant. By expressions (3-3) and (3-4), the *fractional (percentage) error due to x* (with y held constant) *is nothing more than the fractional error in x itself.* Likewise, *the fractional error due to y* (with x held constant) *is the same as the fractional error in y.* The next step is to add these fractional (percentage) errors quadratically, as in expression (3-8). We note that the same kind of argument says that we also add (quadratically) the fractional (percentage) errors when we divide x by y, or vice versa. *For either the product xy or quotient x/y, the fractional (percentage) error of the result is the quadratic sum of the fractional errors in x and y.* In mathematical language, if $z = cxy$ or cx/y, the fractional error is given by

$$\frac{\sigma_z}{z} = \sqrt{\left(\frac{\sigma_x}{x}\right)^2 + \left(\frac{\sigma_y}{y}\right)^2} \qquad (3\text{-}9)$$

Generalization. Often we need more complicated expressions for z as a function of the x and y. Common expressions are linear combinations, $z = ax + by$, and products of powers, $z = cx^N y^M$. These are basically combinations of the operations that we have treated separately: addition or subtraction, multiplication or division by a constant, formation of products or quotients, or raising to a power. Any reader who wants to go into details of how these operations are put together is referred to the references. We present the results without further arguments.

Let us suppose that we have measured independently two quantities, x and y, and want to find the way that errors propagate in a third variable $z(x, y)$, which depends on both x and y. The prescription for finding this is to use expressions (3-2), (3-5), (3-6), and (3-7) for each variable, and then to combine the two sets of expressions quadratically with the expression for two, independent, random measurements. That is, we must combine the sum of squares of effects of the two measurements. The expressions are as follows:

3. Linear combination. If $z = ax + by$, or $z = ax - by$, then

$$\sigma_z = \sqrt{a^2\sigma_x^2 + b^2\sigma_y^2} \qquad (3\text{-}10)$$

4. Product of powers. If $z = cx^N y^M$, the fractional error is given by

$$\frac{\sigma_z}{z} = \sqrt{N^2\left(\frac{\sigma_x}{x}\right)^2 + M^2\left(\frac{\sigma_y}{y}\right)^2} \qquad (3\text{-}11)$$

and the percentage error by

$$\%\text{ Error in } z = \sqrt{N^2(\%\text{ error in } x)^2 + M^2(\%\text{ error in } y)^2} \qquad (3\text{-}12)$$

5. General expression. A more general formula, based on calculus, which includes expressions (3-10), (3-11), and (3-12) as special cases is, if $z = z(x, y)$,

$$\sigma_z = \sqrt{\left(\frac{\partial z}{\partial x}\right)^2 \sigma_x^2 + \left(\frac{\partial z}{\partial y}\right)^2 \sigma_y^2} \qquad (3\text{-}13)$$

When there are more than two measured quantities, you can extend expressions (3-10) through (3-13) by adding more terms under the square-root sign.

3-4 AN IMPORTANT TIP TO SIMPLIFY MATTERS

The material on error analysis in the last section is complicated. Often students and laboratory instructors allow it to become overwhelming. Yet the root mean sum of squares method of calculating errors [expression (3-8)] actually makes things *simpler.* Remember that error estimates usually aren't more precise than ±50%; one significant figure is all that you can expect in your error estimate.

Given this, we will now see that, if one source of error A is appreciably larger than another B, then B has a negligible effect on the final error.

Suppose A is equal to a 2% error and B is half as large. Then, by expression (3-8), the final error estimate is % error $= \sqrt{(1^2 + 2^2)} = 2.24\% \approx 2\%$, since we round errors to single significant figures. *To one significant figure*, the total error is caused completely by the larger source A.

The conclusion is quite general: *a successful error analysis finds the largest source of error*, whether it be systematic or random, rather than attempting to add the effects of many small errors.

3-5 ERROR ANALYSIS WITH YOUR CALCULATOR OR COMPUTER

As an alternative to the methods contained in Sections 3-2 and 3-3, you can trace errors with your calculator or computer. This method is less general and slower, but it is a lot easier to understand!

Measurement of one quantity. We already have seen examples of this method in Sections 3-1 and 3-2. You measure a single quantity x with its error σ_x; you use a formula to calculate some other quantity z. To find the error in z, you simply calculate z twice. The first time you plug x into the formula to get z. The second time you plug in $x + \sigma_x$ and out comes $z + \sigma_z$! The difference between the two numbers is your error in z. Please look again at Sections 3-1 and 3-2 for examples.

Measurement of more than one quantity. In this case, you measure more than one quantity, $x, y, \ldots$, each with its error $\sigma_x, \sigma_y, \ldots$. You treat each quantity $x, y, \ldots$, one at a time, as in the last paragraph. You then get a series of partial errors in z. To get the error in z caused by all variables acting together, take the RS sum, as in expression (3-8)

The computer program *errpro* or *error propagation* (on the enclosed disc) completely automatically does error propagation, with the same algorithm as that used by the calculator method.

Example 3-3 Density of a Sphere

The mass M of a sphere is 1000(1) g and the diameter is $D=8.000(2)$ cm. What is its density? The density ρ of a sphere is given by the expression*

$$\rho = \frac{M}{V} = \frac{M}{\left(\frac{\pi D^3}{6}\right)} \tag{3-14}$$

Solution. By direct substitution in expression (3-14), the density is $\rho = 1000\ \text{g}/[\pi/6 \times (8\ \text{cm})^3] = 3.730\ \text{g/cm}^3$.

Find the error by general formulas. The fractional error in ρ is given by expression (3-11), with $M = 1$ and $N = -3$: $\sigma_\rho/\rho = \sqrt{[(1/1000)^2 + (-3/4000)^2]} = 0.00125$, which gives the final result $\rho = 3.730(5)\ \text{g/cm}^3$.

Find the error with a calculator. We calculate the density with a calculator three times:

1. Use the measured values of M and D. This gives
   $\rho = 1000\ \text{g}/[(\pi/6) \times (8\ \text{cm})^3] = 3.7302\ \text{g/cm}^3$.
2. Use $M + \sigma_M$ and D. This gives
   $\rho = 1001\ \text{g}/[(\pi/6) \times (8\ \text{cm})^3] = 3.7339\ \text{g/cm}^3$.
   The change in z due to the error in M is $3.7339 - 3.7302\ \text{g/cm}^3 = 0.0037\ \text{g/cm}^3$
3. Use M and $D + \sigma_D$. This gives
   $\rho = 1000\ \text{g}/(\pi/6) \times (8.002\ \text{cm})^3 = 3.7274\ \text{g/cm}^3$.
   The change in z due to the error in D is $3.7274 - 3.7302\ \text{g/cm}^3 = -0.0028\ \text{g/cm}^3$.

The error in ρ is the RS sum of the two partial errors: $\sigma = \sqrt{[(0.0037)^2 + (-0.0028)^2]} = 0.0046\ \text{g/cm}^3$, which we round to $\sigma = 0.005\ \text{g/cm}^3$, in agreement with the calculation using formula (3-11).

Find the error automatically. We use the program *errpro* or *error propagation* (on the enclosed disc), with the values $d = 8 \pm 0.002$, $m = 1000 \pm 1$, and

*Note: The formula for the sphere is given in terms of its *diameter* D, not its radius r. A very common error in student laboratories is to measure the diameter of an object and then to substitute it in a formula containing the radius. The tipoff to this mistake is a discrepancy of a factor of 2 (or 4 or 8!) in a calculation. To avoid this problem, *always write expressions in terms of the measured quantity the diameter, not the radius which seldom is measured.*

$F = m/(\text{PI}*d\text{^}3/6)$. The program completely automatically does error analysis by the same algorithm as the calculator method. It computes $\Delta_M = 0.004$, $\Delta_d = -0.003$, with the final result $F = 3.73 \pm 0.005$, in agreement with the calculator.

The final, rounded value for all calculations is $\rho = 3.727 \pm 0.005$ g/cm^3.

Summary

The most important part of error analysis is finding the effect of the largest source of error, whether it be systematic or random. Random errors caused by independent measurements add quadratically: if errors A and B combine to form C, $C = \sqrt{(A^2 + B^2)}$. The length of a random walk of N steps is $\sqrt{N}$ times the length of a single step.

If the error in z is σ, the fractional error in z is σ/z; the percentage error in z is $100\sigma/z$.

In the case of sums, differences, or linear combinations, add (quadratically) the *absolute errors*: If $z = ax + by$, or $z = ax - by$, then

$$\sigma_z = \sqrt{a^2\sigma_x^2 + b^2\sigma_y^2}$$

In the case of products or quotients, add (quadratically) the *fractional* or *percentage* errors: If $z = cxy$ or cx/y, the fractional error is given by

$$\frac{\sigma_z}{z} = \sqrt{\left(\frac{\sigma_x}{x}\right)^2 + \left(\frac{\sigma_y}{y}\right)^2}$$

and the percentage error by

$$\%\text{ Error in } z = \sqrt{(\%\text{error in } x)^2 + (\%\text{ error in } y)^2}$$

If $z = cx^N y^M$, the fractional error is given by

$$\frac{\sigma_z}{z} = \sqrt{N^2\left(\frac{\sigma_x}{x}\right)^2 + M^2\left(\frac{\sigma_y}{y}\right)^2}$$

and the percentage error by

$$\%\text{ Error in } z = \sqrt{N^2(\%\text{ error in } x)^2 + M^2(\%\text{ error in } y)^2}$$

If $z = z(x, y)$,

$$\sigma_z = \sqrt{\left(\frac{\partial z}{\partial x}\right)^2 \sigma_x^2 + \left(\frac{\partial z}{\partial y}\right)^2 \sigma_y^2}$$

Error analysis with a calculator. To find the partial error in a final result z caused by the error in a measurement of a quantity x, calculate the effect of a change with your calculator.

If z depends on more than one variable, do the same for any other partial errors. Then take the root-sum-of-squares of all the partial errors to get the total error.

Error analysis with a computer. The program *errpro* or *error propagation* does a completely automatic error analysis with the same algorithm as that used by the calculator method.

Problems

3-1. Estimate the volume of a sphere $V = \pi d^3/6$ with diameter $d = 1.00(1)$ cm. (a) What is your percentage error in d? (b) In V?

3-2. A cube has the length of one edge $L = 2.001(1)$ in. (a) Give the area and error for one face. (b) What is the percentage error?

3-3. A deep hole has a small step at the bottom, as shown in Fig. 3-5. To measure the height of the step, you put a meter stick in the hole and measure the following data: distance from top of hole to top of step $= 98.0 \pm 0.1$ cm; distance to bottom of hole $= 99.0 \pm 0.1$ cm. There are two sources of error: random errors, which are given, and a systematic error caused by a uniform shrinkage or expansion of ± 1 mm over the length of the meter stick. Make a table showing the values, random error, and systematic errors for each of the three quantities: hole depth, depth to top of step, and step height. Be careful to understand which errors are correlated and which are uncorrelated.

3-4. A ball-bearing factory tests for uniformity of diameter of balls by weighing them. The relative precision of masses is 0.001%. If all the balls have the same density, what is the error in the diameter of a 1-cm ball?

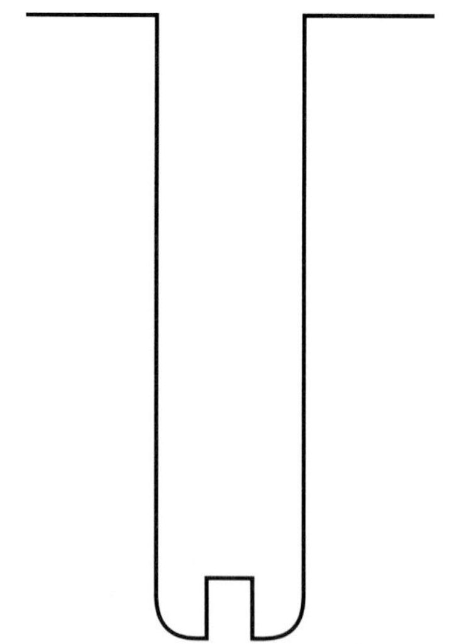

Figure 3-5

3-5. The thickness of a 200-page book (covers excluded) is 3.0(1) cm. Find the (a) absolute error, (b) relative error, and (c) percentage error in the thickness. Find (d) the thickness of a single leaf of the book and its (e) percentage, (f) fractional, and (g) absolute errors.

3-6. A rectangular block of wood measures length $L = 10.0(1)$ cm, width $W = 5.0(1)$ cm, thickness $T = 2.0(1)$ cm, and mass $M = 50.0(1)$ g. Solve for (a) the density of the wood, (b) the error with all sources taken into account, and (c) the error obtained by neglecting all but the largest source. (d) Compare your results from parts (b) and (c) and state your conclusions.

3-7. A brass bar has a diameter $D = 2.50(1)$ cm, a length $L = 30.48(1)$ cm, and a mass $M = 1158.0(1)$ g. Calculate (a) the density of the bar, (b) its fractional error, with all sources taken into account, and (c) the error obtained by neglecting all but the largest source. (d) Compare your results to those from parts (b) and (c) and state your conclusions. The volume of a cylinder is given by $V = \pi D^2 L/4$.

3-8. A freely falling body obeys the equation $y = \frac{1}{2} g t^2$. If $y = 1.000(1)$ m and $t = 0.45(1)$ sec, find (a) the fractional error in y, (b) the fractional error in t, (c) g, and (d) the fractional error in g. (e) Could you neglect any error, or was a complete error analysis needed?

3-9. The frequency of a tank circuit is given by the expression $f = 1/(2\pi\sqrt{LC})$. If L is known to 5% and C to 20%, to what percentage accuracy can you predict the frequency f?

***3-10.** A walker takes steps exactly 1 m long, one each second, but in a random direction. (a) How far will he walk, on the average, in 1 hour? (b) How long will it take him to walk, on the average, 1 km?

3-11. The length, or any linear dimension, of a material at a temperature T in °C is given by the equation $L = L_0(1 + \alpha T)$, where L_0 is the length at 0°C and α is the linear coefficient of expansion. Likewise, the volume of the material is given by the equation $V = V_0(1 + \beta T)$, where V_0 is the volume at 0°C and β is the volume coefficient of expansion. Find the relation between α and β. (Hint: treat αT and βT as small fractional errors and use error analysis, where any shaped volume is given by $V = CL^3$, where C is a constant.)

3-12. Find the volume coefficient of expansion of a Pyrex bulb: $\alpha_{\text{Pyrex}} = 3 \times 10^{-6}{}^\circ\text{C}^{-1}$.

3-13. A constant volume gas thermometer consists of a Pyrex bulb filled with helium gas. The bulb is connected to a manometer, which measures the pressure p at temperature T, given by the equation p = const. $T(K)/V$. For V constant, this becomes $p = p_0 \text{T(K)}/\text{T(K)}_0$, where p_0 is the pressure at 0°C and $T(K) = T + T(K)_0$ is the Kelvin (absolute) temperature. $T(K)_0 = 273$. Put the two equations together and show that $p = p_0[1 + \beta T(°\text{C})]$. (a) Find β. (b) Take into account the volume expansion of the bulb according to the equation $V = V_0(1 + \beta' T)$, where β' is the volume coefficient of expansion of the bulb. Find the observed volume coefficient in terms of β and β'. (c) Find the percent error in β caused by the expansion of the bulb.

4

LINEAR REGRESSION: FITTING A STRAIGHT LINE TO A SET OF POINTS

INTRODUCTION

Linear relations in mathematics, physics, and other fields. Two quantities y and x have a *linear relation* when a graph of y versus x is a straight line. (For an example of a linear relation, see Table 4-1 and Fig. 4-1.) The equation of a straight line is

$$y = a + bx \tag{4-1}$$

The *intercept* a of the straight line is the y-coordinate of the point where the line meets the y-axis ($x = 0$). Two points P_1 and P_2, with coordinates (x_1, y_1) and (x_2, y_2), determine a straight line. The *slope* of the line is the tangent of the angle θ the line forms with the x-axis. The coefficient b* is equal to the difference quotient

$$b = \frac{\Delta y}{\Delta x} = \frac{y_2 - y_1}{x_2 - x_1} \tag{4-2}$$

An example of a linear relation in physics is Hooke's law,

$$F = -kx \tag{4-3}$$

which states that the force F acting on a spring is proportional to the displacement x of the end of the spring from its equilibrium position, which is at $x = 0$.

TABLE 4-1 A Linear Relation: $y = a + bx, a = 1, b = 0.5$

x	0	1	2	3	4	5	6
y	1	1.5	2	2.5	3	3.5	4

*Professor William Hooper has kindly pointed out that the commonly used term "slope" for b is incorrect, except in the case that the x- and y-axes are on the same scale. We shall use the term "difference quotient" instead of "slope" for b.

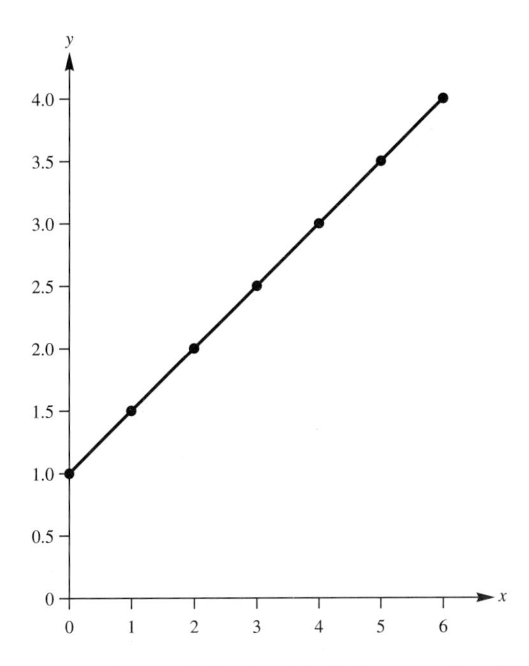

Figure 4-1 Graph of a linear relation (straight line) $y = a + bx$, with intercept $a = 1$ and difference quotient $b = 0.5$.

Because the linear relation is the simplest one between two quantities, you will find it in other fields besides physics. Linear relations are used to predict trends in subjects ranging from population growth to stock prices.

In this chapter, we examine two different ways to fit a linear relation to a set of data. The *graphical* method only requires graph paper, a sharp pencil, a straight edge, and a good eye. The *analytical* method is done most easily with a scientific calculator or, better yet, a computer with built-in linear regression programs.

Later in this book, we will use straight lines to interpret the counting rate of a radioactive isotope or the relation between the length of a pendulum and its period, even though the raw data don't lie on a straight line.

4-1 LINEAR REGRESSION AND ERROR ESTIMATES WITH GRAPH PAPER

Consider a set of measurements of the two variables x and y. For each value of x (which we assume we know exactly), we have a measured value of y. We plot these data on a graph as a set of points (see Table 4-2 and Fig. 4-3 for an example). Since two points determine a straight line, two measurements give us our line. With more than two points, we may not be able to make all the points lie on a straight line. What should we do now?

Graphical method. *Given:* N data points $x_1, y_1; x_2, y_2; \ldots; x_N, y_N$. *To find:* The straight line that is the best fit to the points, the intercept a, the difference quotient b, and its error σ_b.

Method

1. Plot the data

A. *Choose scales.* First, find the *ranges* (smallest and largest values) of x and y. Naturally, your scale for each axis should be wide enough to cover its range of values. Set your scales so that the range of y is equal approximately to the *same* distance covered by the range of x. Your fitted line will then make an angle of about 45° with either axis (slope ≈ 1), which is the most precise way of plotting and reading graphs.

Make your scales convenient both for plotting points and for reading data from the graph. Best are 100, 10, 1, 0.1, 0.01, and so on, units per division; good are scales like $5, 2, \frac{1}{2}, \frac{1}{5}$; fair are 4, 2.5, 0.4, 0.25, and so on. Avoid scales like $7, 3, \frac{1}{3}, \frac{1}{9}$; they are hard to use.

B. *Plot* the points on the graph. Make a sharp dot for accuracy, surrounded with a circular mark for visibility. You can show the error σ in y with *error bars*, straight lines that end at $y + \sigma$ and $y - \sigma$.

2. Find the best straight line

This is easiest with a transparent ruler or triangle so that you can see all the points at once. Lay the straight-edge on the paper, fit it to the points by eye, and draw a straight line. You can do this in two steps, as shown in Fig. 4-2.

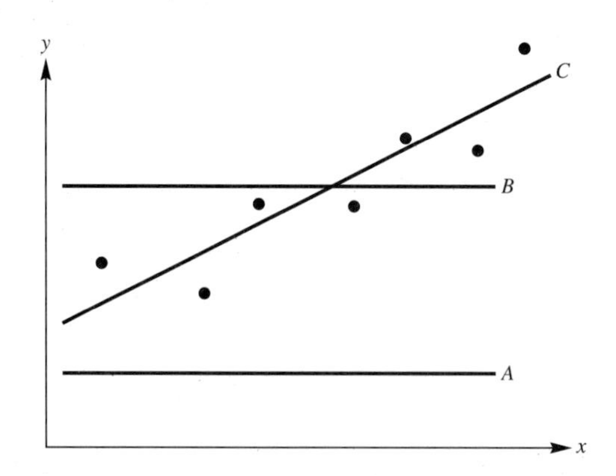

Figure 4-2 Fitting a line to a set of points in two steps. A: Line is too low and has wrong slope. B: Line is at correct height, as it passes through the center of gravity of the points. C: Line is rotated about center for best fit.

a. Move the straight-edge up or down until an equal number of points lie above and below it.
b. Rotate it about the center of the set of points so that step 1 holds equally well for both sets of points, those to the left and those to the right.†

3. Find the difference quotient b and intercept a.

a. The *intercept* a is the point where your line crosses the y-axis.
b. The *difference quotient* b can be found from expression (4-2). Take two points P_1 and P_2 conveniently located on your straight line. Choose easy numbers like $x = 0$ and $x = 10$, and read off the corresponding numbers for y (for an example, see Fig.4-3).

4. Estimate the error in b.

To do this, you construct an error parallelogram, as shown in Fig. 4-3. First, draw two lines *ef* and *gh*, each parallel and displaced by a distance $\pm\sigma$ from your straight line, such that the vertical lines *eg* and *fh* each have a length equal to the

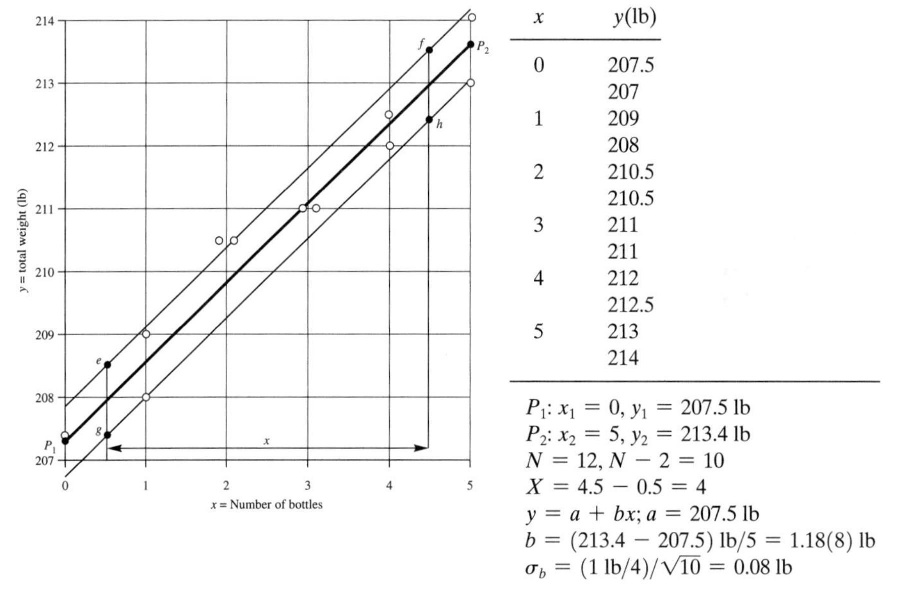

x	y(lb)
0	207.5
	207
1	209
	208
2	210.5
	210.5
3	211
	211
4	212
	212.5
5	213
	214

P_1: $x_1 = 0$, $y_1 = 207.5$ lb
P_2: $x_2 = 5$, $y_2 = 213.4$ lb
$N = 12$, $N - 2 = 10$
$X = 4.5 - 0.5 = 4$
$y = a + bx$; $a = 207.5$ lb
$b = (213.4 - 207.5)$ lb$/5 = 1.18(8)$ lb
$\sigma_b = (1\text{ lb}/4)/\sqrt{10} = 0.08$ lb

Figure 4-3 Example of fitting a straight line to a set of points by the graphical method. The intercept a is at P_1, the intersection of the line with the y-axis. The points P_1 and P_2 on the line determine the difference quotient b. The error parallelogram is chosen to include approximately two-thirds of the points.

†More precisely, make the sum of y-deviations of each point from the line equal to zero in part 1, which is the same as making the line pass through the center of gravity of the points. In part 2, make the sum of y-deviations of each point, multiplied by its x-deviation from the center, equal to zero.

error bars (2σ). If you do not know your error bars, draw *ef* and *gh* such that about two-thirds of your data points lie between the two lines. Then draw two vertical lines *eg* and *fh*, such that the central two-thirds of the data points lie between *eg* and *fh*. Call the horizontal separation between *eg* and *fh* X.

The standard deviation of b is given by the expression

$$\sigma_b = \frac{hf}{X\sqrt{N}} \tag{4-4}$$

if you used error bars; otherwise, use $N-2$ instead of N in Eq. (4-4).

This construction will be correct if the data points are evenly spaced along the x-axis. It can be shown that this method is equivalent to the standard statistical formulas, (4-5), and (4-6), given in the next section.**

Advantages and disadvantages of the graphical approach. The graphical method has one important advantage: you can see what you are doing. If your data don't follow a linear relation, if you have a bad point that falls far off the line, or if you need more data to fill in gaps, a graph tells you right away. Punching numbers in a calculator or computer is a somewhat blind process, which doesn't detect these problems. For this reason, it's always a good idea to plot data, even if you calculate a, b, σ_a, and σ_b electronically.

The graphical method for estimating errors is often not precise enough for physics laboratory data, because the scatter about the line is too small to see clearly. In this case, you can estimate the error by substituting 2σ for hf in expression (4-4), where σ is the y-error σ for single observations. Alternatively, use the calculator programs in Appendix B. The Excel computer programs, furnished with this book, give you the best of both worlds. The programs make the best fit and plot a graph beside it. (See Example 4-2 and Figure 4-4.)

4-2 LINEAR REGRESSION AND ERROR ESTIMATES WITH A CALCULATOR OR COMPUTER

Given: N data points $x_1, y_1; x_2, y_2; \ldots x_N, y_N$. We assume the x-values are exact; the y-values are subject to error.

To find: The straight line that is the best fit to the points, the intercept a with its error σ_a, the difference quotient b, and its error σ_b.

Method: The analytic method of linear regression finds the best fit by minimizing the sum of squares of deviations of y values from the fitted straight line. For this reason, it is called the *method of least squares.* The results of this method

A similar construction was given previously by J. L. Safko, *Amer. J. Physics* **33, 379–382 (1965). For more details, including a graphical method for finding the error in the intercept a, see W. Lichten, *Am. J. Phys.* **57**(12), 1112–1115. (1985)

are in the form of formulas which are derived by calculus in the references. See also the problems at the end of this chapter. The expressions are as follows:

$$b = \frac{\Sigma(x - \overline{x})(y - \overline{y})}{N\sigma_x^2}, \qquad a = \overline{y} - b\overline{x} \tag{4-5}$$

$$\sigma_b = \frac{\sigma}{\sigma_x\sqrt{N}}, \qquad \sigma_a = \sigma_b\sqrt{\frac{\Sigma x^2}{N}} \tag{4-6}$$

where $\overline{x}, \overline{y}$ are the means of the x and y values, respectively, as given by expression (1-1), σ_x is the standard deviation of the x values as given by expression (2-2), and σ is the estimated error in y. In case we find σ from scatter, we use the expression $\sigma = \sqrt{\frac{\Sigma d^2}{(N-2)}}$, where $d = y - (a + bx)$ is the y-deviation of a point (x, y) from the fitted straight line $y = a + bx$.

Direct calculation of the quantities a, b, σ_a, and σ_b is complicated, tiresome, fraught with the possibility of error, and is in Appendix C. The graphical method of the last section avoids much of this complexity. It is also easier to do these calculations with a preprogrammed calculator or computer. At the time of writing of this book, calculators that do linear regression (find a and b) are available, but no calculators find the errors σ_a and σ_b. Algorithms and programs to find these quantities with calculators and computers are given in Appendix B. Each program has a simple example that you can use to test yourself, to make sure you are using the program correctly. A computer spreadsheet program is the easiest way to do line fits. Most spreadsheet programs will find difference quotient b, intercept a and the errors, but take some time to learn the procedures. The disc enclosed with this book has Excel programs that require only entering the raw data.

Example 4-1 Weighing Bottles on a Bathroom Scale

Your problem is to weigh a filled bottle (weight $\approx$ 1 lb). All you have for weighing are a set of bathroom scales, precise to about one-half pound for a single measurement. How are you to weigh the bottle?

Your ingenious solution to this problem is to get several bottles, stand on the scales and weigh yourself with x bottles. You start with $x = 0$, then $x = 1, 2$, etc. until you hold all the bottles. To be extra careful, you unload one bottle at a time and repeat the measurement. You then plot the data, your total weight plus that of x bottles. You fit the data with a straight line. The difference quotient b of the line gives the weight per bottle, i.e., the weight of one bottle. The intercept gives your weight.

Figure 4-3 shows the data and graphical analysis. The final results are your weight (intercept a) is 207.5 lb† and the weight per bottle (difference quotient b) is 1.18(8) lb.

The details of the analytic computation are shown in Appendix C. The results are intercept: $a = 207.4(2)$ lb and difference quotient: $b = 1.23(8)$ lb. The graphical and analytical results agree well, as the differences are less than the errors. The quantity $\sigma = 0.46$ lb is a good estimate for the error of a single measurement of the total weight y.

†See footnote on page 46.

Example 4-2 Measurement of the Force Constant of a Spring

Bob hangs a mass from a spring and measures the downward displacement y in cm as a function of x in grams. Bob measures y to the nearest $\frac{1}{2}$ mm (0.05 cm) and takes the average for x values running from 0 to 40 cm and back again. If the spring obeyed Hooke's law, there would be a linear relation between y and x:

$$y = bx \tag{4-7}$$

Bob enters the observations in the EXCEL program *FitLine* (in the enclosed disc), which graphs the data points, fits the best line $y = a + bx$, and gives a and b with their errors (Fig. 4-4a). The straight-line fit to Eq. (4-11) is not good. The 0 and 10 cm points deviate by about a centimeter. In both cases, the deviations are far in excess of the least count of $\frac{1}{2}$ mm. The reason for this behavior is that the spring is closely wound and is actually under tension with no load.

Bob discards the zero load point in Fig. 4-4b. The points now fit a line with no noticeable deviations. The error in b, and therefore in the force constant, is cut by a factor of 20. (Another quality measure, the standard deviation S_y of the data points from the line, is cut by a factor of 30 to 0.03, which is consistent with the least count of 0.05 cm.) Thanks to the graph, Bob made a major improvement in the measurement of the force constant. The saying *"One picture is worth a thousand words" is still true in the computer age.*

Enter Line Fit Data:	
Xi	Yi
Added mass (g	Stretch (cm)
0	0
10	3.6
20	9.425
30	15.35
40	21.225

EQUAL WEIGHT LINE FIT			
Slope B =	0.542	±	0.02656909
Intercept A =	-0.92	±	0.65080719
SumSqDev=	2.11775	StdDev Sy=	0.84018847

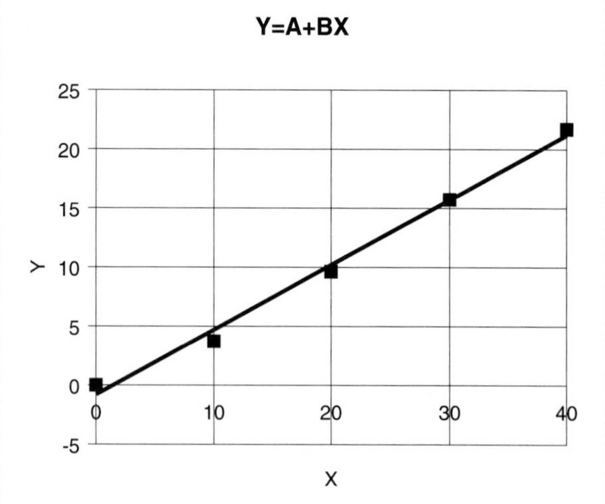

Y=A+BX

Figure 4-4a

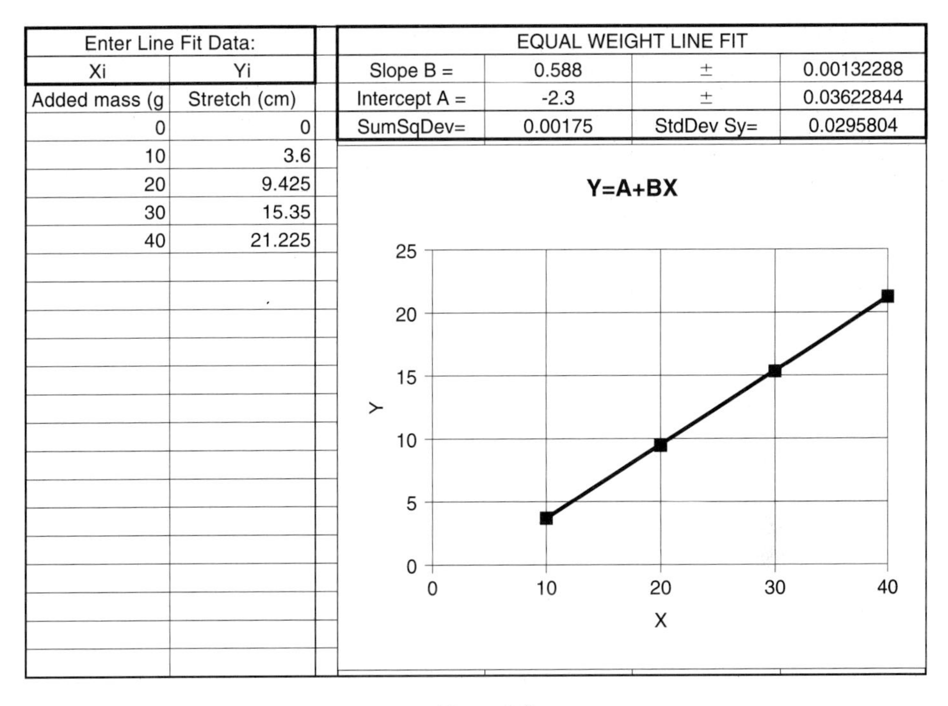

Enter Line Fit Data:	
Xi	Yi
Added mass (g	Stretch (cm)
0	0
10	3.6
20	9.425
30	15.35
40	21.225

EQUAL WEIGHT LINE FIT			
Slope B =	0.588	±	0.00132288
Intercept A =	-2.3	±	0.03622844
SumSqDev=	0.00175	StdDev Sy=	0.0295804

Figure 4-4b

Example 4-3 Determination of the Absolute Zero of Temperature

Bridget finds the absolute zero of temperature with a constant-volume helium thermometer. Helium is nearly a perfect gas, which has pressure proportional to absolute temperature at constant volume. At two temperatures, 99.9°C and 1.0°C, the absolute pressures, in millimeters of mercury, are 742 and 545 mm, respectively. Bridget's assignment is to

a. find the absolute zero of temperature on the Celsius scale.
b. find the statistical error, with an assumed least count of 1 mm for the pressure measurement and random errors.
c. find the systematic error which would be caused by a uniform shift in pressure of 1 mm for both readings, such as would occur for an inaccurate barometer measurement.

Solution. Bridget plots temperature versus absolute pressure, the sum of barometric and gauge pressures, as shown in Figure 4-5. To find the extrapolated temperature at zero pressure, she uses a linear regression program, such as *FitLine* (see Appendix B).

a. The result is $a = -272.6$°C with zero error due to scatter, since two points determine a straight line exactly.
b. For the statistical error, she uses the calculator method or the cT error analysis program, with an error of the least count of 1 mm. For the calculator method, she

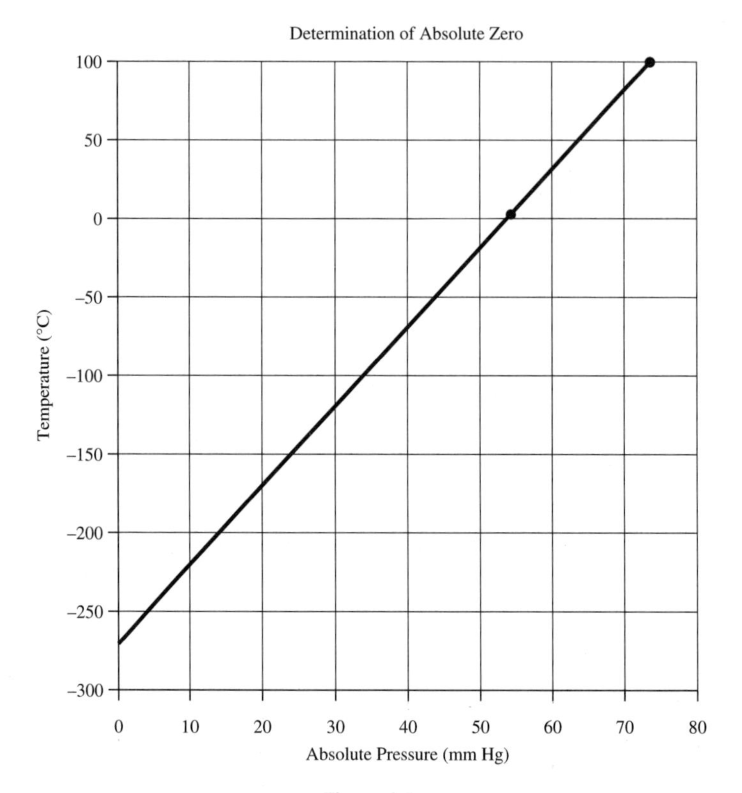

Figure 4-5

changes each pressure in turn by 1 mm and repeats the calculation for absolute zero with the regression program. The shifts are 1.9 and 1.4 mm, which give a root square sum of 2.4°C, which is the statistical error. She is pleased to find absolute zero at −273(2)°C, in agreement with the accepted value.

c. For the systematic error, she changes both pressures by 1 mm and reenters the linear regression program. The result is a shift of −0.5°C, which is her estimate of systematic error due to a pressure shift of +1 mm (Fig. 4-5).

4-3 OTHER TOPICS: CORRELATION COEFFICIENT *r*, GROUPED OR WEIGHTED DATA, NONLINEAR FUNCTIONS

The correlation coefficient *r*. We note, for reference purposes only, the correlation coefficient r, which is defined for a set of data points (x, y) by the equation

$$r = \frac{b\sigma_x}{\sigma_y} \tag{4-8}$$

where σ_x, σ_y are the standard deviations of the x and y values, respectively, as defined in Eq. (2-2). This coefficient r is always between -1 and $+1$. The two values ± 1 represent perfect fits of the data to straight lines, with no scatter. The correlation coefficient is used mainly by social scientists, whose data contain more scatter than those of physical scientists. Typical data in the elementary physics laboratory have r equal or greater than 0.99 in magnitude; thus, like a professor who gives nothing but A+ for a grade, r is not a useful measure of the quality of data.

However, r cuts the amount of button pushing in computing linear regression errors (see Appendix B). In terms of r, expression (4-6) becomes

$$\sigma_b = b\sqrt{\frac{\left(\frac{1}{r}\right)^2 - 1}{N - 2}} = \frac{b\tan\left[\cos^{-1}(r)\right]}{\sqrt{N - 2}} \tag{4-6'}$$

$$\sigma_a = \sigma_b\sqrt{(\Sigma x^2/N)}$$

Linear regression with grouped data. This section discusses methods for handling data with large numbers of points. The related technique of handling weighted data, most useful in calculation of half-lives of radioactive isotopes, is discussed in the next chapter.

Graphical method. The graphical method (Section 4-1) of fitting a straight line to a set of points easily handles grouped data. Where there are n points with the same values of x and y, merely make a cluster of n dots on your graph close by the point (x, y). The construction of the error parallelogram and the rest of the computation go exactly as before.

Analytic (Calculator) method. Some calculators are preprogrammed to allow you to put in grouped or weighted data directly and thus present no problem. With other calculators or computers, expressions (4-5) and (4-6) must be modified. The algorithms and programs for these calculators are rewritten in Appendix B to handle weighted data for two important cases: the RC time constant and radioactivity.

Expressions for linear regression with grouped data. The expressions for data, where each point (x, y) occurs f times, are as follows:

$$b = \frac{\Sigma f(x - \overline{x})(y - \overline{y})}{\Sigma f\sigma_x^2}, \qquad a = \overline{y} - b\overline{x} \tag{4-9}$$

$$\sigma_b = \frac{\sigma}{\sigma_x}\sqrt{\frac{1}{N - 2}}, \qquad \sigma_a = \sigma_b\sqrt{\frac{\Sigma f x^2}{N}}, \tag{4-10}$$

where $\overline{x}$, $\overline{y}$ are the means of the x- and y-values, respectively, as given by expression (2-10), σ_x is the standard deviation of the x-values as given by expression (2-11), $d = y - (a + bx)$ is the y-deviation of a point (x, y) from the fitted straight line $y = a + bx$, N is the total number of points, given by expression (2-9): $N = \Sigma f$, and σ is the y-error for a single data point by the scatter method:

$$\sigma = \sqrt{\frac{\Sigma f d^2}{N}}.$$

Some calculators are wired with a frequency button to handle expressions (4-9) and (4-10) automatically. In these cases, expression (4-6′) still holds for error computation. (See Appendix B.)

Linear regression with nonlinear functions. You can stretch linear regression techniques to fit functions that don't lie on a straight line. To do so, you convert the nonlinear function into a linear one. A common example is radioactivity, which involves logarithmic functions. This is discussed in Chapter 5. Other examples are pendulums and springs, which involve square roots.

A pendulum, a small bob hanging from a string of length L, has a period T (time for one back-and-forth swing), given by the expression

$$T = 2\pi\sqrt{\frac{L}{g}} \tag{4-11}$$

where $g \approx 980$ cm/sec^2 is the acceleration of gravity. We can find g by plotting T^2 along the y-axis and L along x. This gives a straight line with slope b equal to $4\pi^2/g$. The acceleration of gravity is $g = 4\pi^2/b$.

Another example of a nonlinear function is given by a vertical spring of force constant k [see expression (4-3)], with a mass M hanging at the bottom end. The period of oscillation is given by

$$T = 2\pi\sqrt{\frac{M}{k}} \tag{4-12}$$

By plotting the quantity T^2 versus M, we get a straight line with difference quotient $b = 4\pi^2/k$. The force constant is given by $k = 4\pi^2/b$.

Expressions 4-11 and 4-12, when rearranged as discussed in the past two paragraphs, predict straight lines passing through the origin. At times, experimental results may give a nonzero intercept. This intercept sometimes has physical meaning. For example, we can show that the intercept in the case of the spring is $a = bm/3$, where m is the mass of the spring.

Summary

Least-Squares Fit of a Set of Data Points to a Straight Line y 5 a 1 bx

1. The *graphical method* gives both a and b. Construction of an error parallelogram also gives the error in b.
2. The *analytical solution* is

$$b = \frac{\Sigma(x - \overline{x})(y - \overline{y})}{N\sigma_x^2}, \qquad a = \overline{y} - b\overline{x}$$

$$\sigma_b = \frac{1}{\sigma_x}\sqrt{\frac{\Sigma d^2}{N(N-2)}}, \qquad \sigma_a = \sigma_b\sqrt{\frac{\Sigma x^2}{N}}$$

This can be applied to nonlinear functions and to grouped data.

Problems

Note: Many of these problems can be done most easily with a combination linear regression and graphing program, such as *FitLine*, which is on the disc in the back of this book.

4-1. A rider moves at constant speed along an air track. Find an expression of the form $x = x_0 + vt$ from the data $t = 2.00(5)$ sec, $x = 50.00$ cm, and $t = 3.45(5)$ sec at $x = 100.00$ cm. (a) What is v? (b) What is its error? (c) At what time did the rider pass the $x = 0$ mark? (d) At what value of x was the rider at $t = 0$?

4-2. A vertical spring with an added mass m has a length y, as shown in Table 4-2.

TABLE 4-2 Effect of Added Mass on Length of a Spring.

Mass m (kg)	0	1	2	3	4	5	6
Length y (m)	0.285	0.320	0.392	0.463	0.534	0.604	0.677
Mass m (kg)	7	8	9	10			
Length y (m)	0.748	0.819	0.891	0.962			

Plot the data and fit it with a straight line of the form $y = a + bm$. (a) Find the difference quotient b and its error. (b) Inspect the graph and decide if the data can be improved. How? (c) Repeat part (a) with the improved data.

4-3. Table 4-3 contains four data sets, each consisting of eleven (x, y) pairs. For the first three data sets, the x values are the same, and they are listed only once. For each data set, use a linear regression program to fit a straight line of the form $y = a + bx$. Find the values of a, b, σ_a, and σ_b; then plot a graph for each of the four data sets. Compare the graphical and analytical results. What do you conclude about the importance of first making a graph before punching your data into a calculator?

TABLE 4-3 Four Data Sets, Each Comprising Eleven (x, y) Pairs.

Data set:	1–3	1	2	3	4	4
Variable:	x	y	y	y	x	y
Obs. no. 1 :	10.0	8.04	9.14	7.46 :	8.0	6.58
2 :	8.0	6.95	8.14	6.77 :	8.0	5.76
3 :	13.0	7.58	8.74	12.74 :	8.0	7.71
4 :	9.0	8.81	8.77	7.11 :	8.0	8.84
5 :	11.0	8.33	9.26	7.81 :	8.0	8.47
6 :	14.0	9.96	8.10	8.84 :	8.0	7.04
7 :	6.0	7.24	6.13	6.08 :	8.0	5.25
8 :	4.0	4.26	3.10	5.39 :	19.0	12.50
9 :	12.0	10.84	9.13	8.15 :	8.0	5.56
10 :	7.0	4.82	7.26	6.42 :	8.0	7.91
11 :	5.0	5.68	4.74	5.73 :	8.0	6.89

Reprinted with permission, American Statistical Association, from "Graphs in Statistical Analysis," by F. J. Anscombe, *The American Statistician*, Vol. 27, No. 1, pp. 17–21.

4-4. The displacement versus time function for a body released at rest and in uniformly accelerated motion is $x = \frac{1}{2}at^2$. A plot of x versus t^2 gives a straight line with difference quotient $b = a/2$. Table 4-4 gives the time displacement relation for a rider on an air track. Find the (a) acceleration a and (b) its error.

TABLE 4-4 Displacement Versus Time for an Air-Track Rider.

Time, t (sec)	0	1	2	3	4	5	6
Displacement, x (cm)	0	2.1	7.3	19.3	33.5	48.0	71.0

4-5. The average velocity of a body between times t_n and t_{n+1} is given by the expression

$$\overline{v} = \frac{x_{n+1} - x_n}{t_{n+1} - t_n}$$

In uniformly accelerated motion, a plot of $\overline{v}$ versus t gives a straight line with difference quotient a. Use the data in Table 4-4 to find such a relation. For $\overline{v}$, simply take the differences between neighboring values of x. Find the difference quotient a and its error. Compare with the answer to Problem 4-4.

TABLE 4-5 Average Heights and Weights of 20- to 24-Year-Olds.

Height (in.) W	58	59	60	61	62	63	64	65	66	67	68	69	70	71	72
Weight (lb) W	105	110	112	116	120	124	127	130	133	137	141	146	149	155	157
Height (in.) M	62	63	64	65	66	67	68	69	70	71	72	73	74	75	76
Weight (lb) M	130	136	139	143	148	153	157	163	167	171	176	182	187	193	198

4-6. The average heights H and weight W of fully clothed U.S. men and women, 20 to 24 years old, are given in Table 4-5. For either, try fitting a straight line of the form $W = a + bH$ to the data. Give your results for a and b and the error in b. If time permits, try investigating further. Try alternatively a quadratic fit of the form $W = d + cH^2$, give c and its error.

Compare the two fits within the range of data. Does either seem a better fit, by appearance of the graph or by the errors in the coefficients? Next, try for how the fitted equation works outside the range of data, (i.e., how it *extrapolates*). See what your fitted function predicts for the weight of a 30-in.-high midget. Which makes more sense, the straight line or the quadratic fit? This exercise illustrates that a good fit may only work within the range of the data; it is dangerous to predict outside the range of the data on the basis of statistics alone.

4-7. The resonant frequency of a tank circuit is given by the expression $f = 1/(2\pi\sqrt{LC})$, where L is the inductance (in henries) and C the capacitance (in farads). In a given circuit, an extra capacitance C' is added and the frequency f measured as a function of C', as shown in Table 4-6. Use the relationship $1/f^2 = L(C + C')/4\pi^2$, where f, C, and C' are in the units in this table. This is the equation of a straight line in $1/f^2$ versus C', with difference quotient $b = L/4\pi^2$ and intercept $LC/4\pi^2$. Solve for L and C and their errors.

TABLE 4-6 C' Versus f for a Resonant Circuit.

Capacitance, C' (μF)	0	1	2	3	4	5
Frequency, f (kHz)	0.700	0.570	0.505	0.440	0.415	0.380

4-8. The acceleration of gravity can be measured remarkably accurately with this simple equipment: a screen door spring, a set of weights, a meter stick, and a stopwatch. First, measure and plot the displacement of the spring end as a function of added mass. Hooke's law says the resulting straight line has a difference quotient $b = g/k$, where g is the acceleration of gravity and k the force constant. Then measure the period of oscillation of the spring as a function of added mass. The plot of m versus T^2 gives a difference quotient $b = k/4\pi^2$. Combining these two coefficients gives g.

Table 4-7 gives the position y of the end of a spring and the added mass M. Plot y as a function of M. Find the difference quotient $b = g/k$ and its error.

TABLE 4-7 Stretch of a Spring y Versus Added Mass M.

Mass, M (kg)	0	1	2	4	4.5
Position, y (m)	0.0179	0.0467	0.078	0.1416	0.1572

***4-9.** Table 4-8 gives the period of oscillation T and mass added to the same spring as in Problem 4-8. For M versus T^2, find (a) the difference quotient and error of the fitted straight line $b = k/4\pi^2$. (b) Combine the answers to Problems 4-8 and 4-9a to solve for the acceleration of gravity g and its error.

TABLE 4-8 Period of a Spring and Added Mass.

Period, T (s)	0.445	0.52	0.63	0.723	0.844
Added mass, M (kg)	0.5	1.0	2.0	3.0	4.5

4-10. Table 4-9 gives the period of oscillation T for the mass m added to a spring. (a) Fit a straight line to T^2 versus m. Find both the value and errors for the intercept a and difference quotient b. (b) Find the mass of the spring and its error. (c) Find the force constant k and its error.

TABLE 4-9 Period of a Spring and Added Mass.

Period, T (sec)	0.52	0.71	0.86	1.00	1.10
Added mass (g)	10	20	30	40	50

4-11. If in example (4-3), we assume a pressure error of half the least count ($\frac{1}{2}$ mm), what are the errors at the upper temperature, lower temperature, and a combination of both?

4-12. *Determination of the temperature of absolute zero with a lightbulb.* The resistance of tungsten is proportional to the absolute temperature. Jill measures the resistance of a flashlight bulb as a function of its temperature, which she determines from its color. Her results are in Table 4-10.

TABLE 4-10 Jill's Temperature-Resistance Data.

Resistance (ohms)	Color	Temperature (°C)
15.4	Bright red	850–950
18.7	Yellowish-red	1050–1150
20.6	Incipient white	1250–1350
23.0	White	1450–

(a) Fit her data for the resistance (x)-temperature (y) relation with a straight line. Use the intercept to find T(°C) of absolute zero. (b) Find the error in T of absolute zero. (c) Find the standard error of Jill's temperature measurements. [Hint: see Example (4-3).]

5

USING TRIGONOMETRIC AND EXPONENTIAL FUNCTIONS IN THE LABORATORY

Introduction

Trigonometric (Gr. the science of triangles) functions are useful in mechanics and optics. Exponential functions, with the related logarithms, deal with radioactive decay, transients in LR and RC circuits, and cell growth. Logarithms are a useful aid in finding the power law relation between two variables. Logarithmic functions also occur frequently in electricity and magnetism. In chemistry, pH is defined in terms of logarithms, and exponential functions occur in the Arrhenius law of reaction rates.

This chapter begins with a review of the basic mathematics of exponential functions and then moves on to apply these functions in the laboratory. It shows how to use these functions by means of the calculator and computer programs that are in Appendix B at the end of the book.

5-1 TRIGONOMETRY

Angles: radians, decimal degrees or degrees, minutes, and seconds. The laws of physics treat angles in *radian* measure. Yet laboratory instrument scales are marked in *degrees, minutes*, and *seconds*. Most calculators turn on in *decimal degree* mode. This section deals with these three angular measures and how to get back and forth among them.

Figure 5-1 illustrates the definition of an angle, θ, in *radians*, in terms of the arc length AB and the radius R of a circle:

$$\theta = \overline{AB}/R \tag{5-1}$$

One revolution subtends an angle $\theta = 2\pi$ radians. Angles also are measured in *degrees*, where $1° = 1/360$ rev $= 2\pi/360 = 0.017\ 453\ 292$ radians. The *minute* is defined by the equation $1' = 1/60°$. The *second* is given by the equation $1'' = 1/60'$. Table 5-1 gives conversions from each of these units to the others.

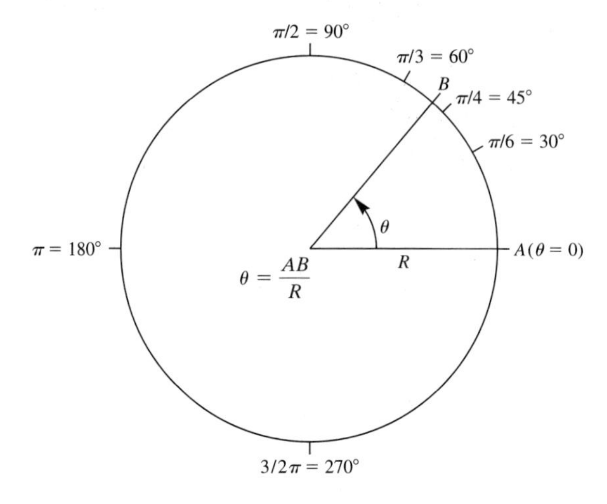

Figure 5-1 Definition of angle in radians: $\theta = \overline{AB}/R$, where $\overline{AB}$ is the arc length subtended by the angle θ and R is the radius of the circle. Several other common angles are shown in radian and degree measure.

Trigonometric functions: sine, cosine, tangent, arc sine, arc cosine, and arc tangent. Given a right triangle with an angle θ, opposite side a, adjacent side b, and hypotenuse c (opposite the right angle), the trigonometric functions are defined as

$$\begin{aligned} \sin(\theta) &= \frac{\text{opp}}{\text{hyp}} = \frac{a}{c} \\ \cos(\theta) &= \frac{\text{adj}}{\text{hyp}} = \frac{b}{c} \\ \tan(\theta) &= \frac{\text{opp}}{\text{adj}} = \frac{a}{b} \end{aligned} \tag{5-2}$$

TABLE 5-1 The Angle θ in Different Units and Its Trigonometric Functions.

Decimal Degrees	Degrees, Min, Sec	Radians	Sin(θ)	Cos(θ)	Tan(θ)	Comments
0.	0° 0′ 0″	0.	0.	1.0	0.	Zero
0.0002778	0° 0′ 1″	0.0000048	0.0000048	1.000000	0.0000048	1 second
0.0166667	0° 1′ 0″	0.0002909	0.0002909	1.000000	0.0002909	1 minute
1.0	1° 0′ 0″	0.0174533	0.0174524	0.9998477	0.0174551	1 degree
57.29578	57° 17′ 44″	1	0.841471	0.5403023	1.5574077	1 radian
30	30° 0′ 0″	0.5235988	0.5	0.8660254	0.5773503	30° = π/6 rad
45	45° 0′ 0″	0.7853982	0.7071068	0.7071068	1.000000	45° = π/4 rad
60	60° 0′ 0″	1.0471976	0.8660254	0.5	1.7320508	60° = π/3 rad
90	90° 0′ 0″	1.5707963	1.0000000	0.000000	∞	90° = π/2 rad = right angle

and
$$\cot(\theta) = \frac{\text{adj}}{\text{opp}} = \frac{b}{a}$$

The *inverse* trigonometric functions are defined by the equations

$$\theta = \text{arc sin}\,\frac{a}{c} = \sin^{-1}\frac{a}{c} = \text{INV sin}\,\frac{a}{c}$$

$$\theta = \text{arc cos}\,\frac{b}{c} = \cos^{-1}\frac{b}{c} = \text{INV cos}\,\frac{b}{c} \qquad (5\text{-}3)$$

$$\theta = \text{arc tan}\,\frac{a}{b} = \tan^{-1}\frac{a}{b} = \text{INV tan}\,\frac{a}{b}$$

and
$$\theta = \text{arc cot}\,\frac{b}{a} = \cot^{-1}\frac{b}{a} = \text{INV cot}\,\frac{b}{a}$$

With the exception of cot and arc cot, these functions are available on scientific calculators.

It follows, by combining Eqs. (5-2) and (5-3), that an angle is the inverse of any of its own trigonometric functions. For example, $\theta = \text{INV}\ \sin(\sin(\theta))$. This equation is useful for converting an angle from one unit to another by means of a pocket calculator, as is discussed in Appendix B.

Table 5-1 gives the values of trigonometric functions and converts from one set of units to another for several common angles. Use this table as a check on your ability to convert units with your calculator or computer.

5-2 APPROXIMATIONS FOR TRIGONOMETRIC FUNCTIONS

We note, for reference purposes, some results of differential calculus and the theory of power series for trigonometric functions. These results are useful for estimating errors without scientific calculators, or as an alternative method of estimating trigonometric functions.

Power series for trigonometric functions. For small value of θ, the following power series hold for θ in radian measure:

$$\sin(\theta) = \cos\left(\frac{\pi}{2} - \theta\right) = \theta - \frac{\theta^3}{3!} + \cdots \qquad (5\text{-}4)$$

$$\cos(\theta) = \sin\left(\frac{\pi}{2} - \theta\right) = 1 - \frac{\theta^2}{2!} + \frac{\theta^4}{4!} + \cdots \qquad (5\text{-}5)$$

$$\tan(\theta) = \cot\left(\frac{\pi}{2} - \theta\right) = \theta + \frac{\theta^3}{3} + \cdots \qquad (5\text{-}6)$$

Also, by applying differential calculus, we have, for x small,

$$\sin(\theta + x) \approx \sin(\theta) + x\cos(\theta) + \cdots \tag{5-7}$$

$$\cos(\theta + x) \approx \cos(\theta) - x\sin(\theta) + \cdots \tag{5-8}$$

$$\tan(\theta + x) \approx \tan(\theta) + \frac{x}{\cos^2(\theta)} + \cdots \tag{5-9}$$

Example 5-1

The acceleration of a body moving freely on a plane inclined at an angle θ with the horizontal is

$$a = g\sin(\theta) \tag{5-10}$$

An air track 2 m long is on a level table. A block 5 cm thick elevates one end of the air track. Calculate the acceleration of a rider on the air track, given that $g = 980$ cm/sec^2.

Solution. The air track forms the hypotenuse of a triangle with $c = 200$ cm and $a = 5$ cm. Then $\sin\theta = 5$ cm/200 cm $= 0.04$ and, by Eq. (5-2), we have $a = 0.04 \times 980$ cm/sec^2 = 39.2 cm/sec^2.

Example 5-2

The equation for a diffraction grating is

$$d\sin\theta = \lambda$$

where d is the grating spacing, θ is the angle between zero and first orders, and λ is the wavelength. A student uses Na D light with $\lambda = 589$ nm (1 nm $= 10^{-9}$ m) and measures the following angles on a spectrometer, which she reads to the nearest minute of arc:

Zero order: $342^\circ\ 11(1)'$

First order: $340^\circ\ 29(1)'$

where the parenthesized quantities show the $1'$ error of measurement. Find the grating spacing d and estimate the error.

Solution. We take the difference between the two given angles:

$$\theta = 342^\circ\ 11(1)' - 340^\circ\ 29(1)'$$

$$= 341^\circ\ 71(1)' - 340^\circ\ 29(1)' = 1^\circ\ 42(1.4)'$$

since the error of the difference of two numbers is $\sqrt{2}$ times the error of one of the two quantities [see expression (3-1)]. We convert this result to decimal degrees: $1^\circ\ 42(1.4)' = 1^\circ + 42/60^\circ = 1.700^\circ$, and the error becomes $1.4' = 1.4/60^\circ = 0.023^\circ$. We combine these numbers as follows: $\theta = 1.700(23)^\circ$ and find the grating spacing:

$$d = \lambda/\sin\theta = \frac{589 \times 10^{-9}\ \text{m}}{\sin[1.700(23)]^\circ} = \frac{5.89 \times 10^{-9}\ \text{m}}{0.02967} = 19{,}850\ \text{nm} = 0.01985\ \text{mm}$$

We now find the error in d by the calculator method (see Section 3-2) by repeating the previous calculation, but increasing θ by its error: $d + \sigma_d = 589$ nm/sin(1.723°) = 19,589 nm, a change of 261 nm, which we round to 300 nm. The final result is d = 19,850(300) nm = 0.019 85(30) mm.

Time conversion. There is a handy parallel between angle and time units:

$$1° = 60' = 3600'' \tag{5-11}$$

and

$$1 \text{ hr} = 60 \text{ min} = 3600 \text{ sec}$$

Hours and degrees are analogous, as are minutes of time and minutes of arc; likewise, seconds of time and seconds of arc. Thus you can use the same calculator or computer program to convert decimal hours to hours, minutes, and seconds, and vice versa.

Example 5-3

A radioactivity experiment has a background count of $N = 486$ counts in a time $T = 1$ hr, 15 min, and 12 sec. What is the counting rate $R = N/T$ in (a) counts per hour, (b) counts per minute, (c) counts per second, and (d) counts per 10 min?

Solution. To answer part (a), we find the time in decimal hours by using trigonometric conversion: $1° \, 15' \, 12'' = (1 + 15/60 + 12/3600)° = 1.2533°$, which can be done in one step on some calculators (see Appendix B). Likewise, we have 1 hr 15 min 12 sec = 1.2533 hr.
(a) The counting rate is $R = N/T = 486/1.2533$ hr = 388 counts/hr.
(b) We convert to counts per minute: $R = 388$ counts/hr × 1 hr/60 min = 6.46 counts/min.
(c) We convert to counts per second: $R = 6.46$ counts/min × 1 min/60 sec = 0.108 counts/sec.
(d) The number of counts in 10 min is $N = R \times T = 6.46$ counts/min × 10 min = 64.6.

5-3 PROPERTIES OF SINUSOIDAL WAVEFORMS

Trigonometric functions represent sinusoidal waveforms, which commonly occur in the laboratory in acoustics, wave motion, ac circuits, and physical optics. We summarize the properties of waves as occur in trigonometric representations. Given a displacement y as a function of time,

$$y = A \sin(\omega t - \delta) \tag{5-12}$$

where A is the *amplitude* (sometimes called the peak amplitude), ω is the *angular velocity*, and δ is the *phase shift*. We list some standard relations (see Fig. 5-2):

1. Amplitudes
   a. Root-mean-square (RMS) (also called ac amplitude): $A_{RMS} = A/\sqrt{2} = 0.707$ A.
   b. Peak-to-peak = $A_{PP} = 2$ A.
2. Frequencies and period
   a. Frequency $f = \omega/2\pi$, the number of complete oscillations per second, in units of sec^{-1} or hertz (Hz).
   b. Period $T = 1/f = 2\pi/\omega$, the time it takes for one complete oscillation.

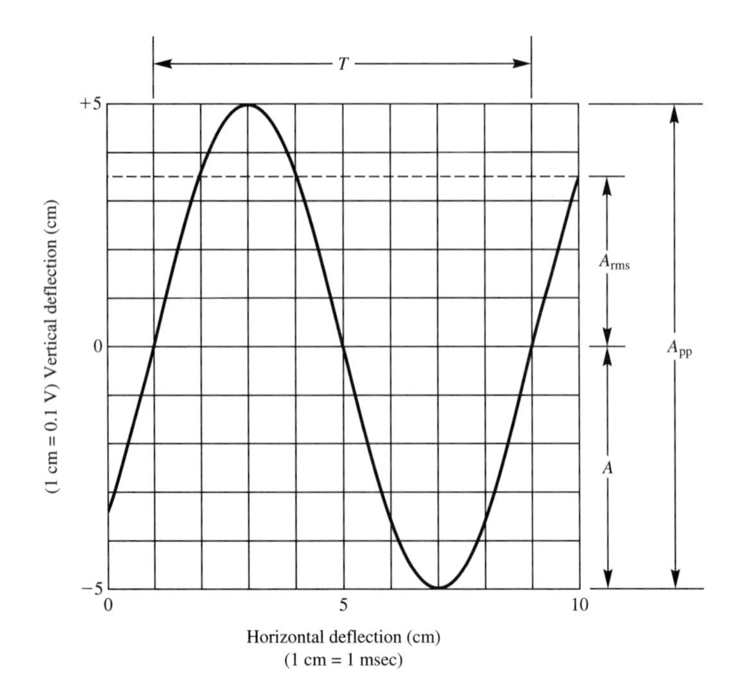

Figure 5-2 Oscilloscope trace of a sine wave. T = period, A = amplitude, A_{RMS} = RMS (ac) amplitude, A_{PP} = peak-to-peak amplitude.

Example 5-4 Measuring Frequency and Phase Shift of a Waveform from an Oscilloscope

Given the waveform in Fig. 5-2, find the period, frequency, angular velocity, phase shift, (peak) amplitude. RMS amplitude, and peak-to-peak amplitude.

Solution. The period is given by the displacement between two identical points on the waveform. We take two points where the waveform crosses the axis, at 1 cm and at 9 cm. (This is more precise than taking the locations of the peaks, which have a poorly defined x-coordinate.)

The period is $T = (9\text{ cm} - 1\text{ cm}) \times (1\text{ ms/cm}) = 8\text{ ms} = 0.008\text{ sec}$
The frequency is given by $f = 1/T = 125\text{ sec}^{-1} = 125\text{ Hz}$
The angular velocity is $\omega = 2\pi/T = 785\text{ sec}^{-1}$
The phase shift can be found by setting $\sin(\omega t - \delta) = 0$ in Eq. (5-12), which occurs at the crossing of the axis at $x = 1$ cm or $t = 0.001$ sec. By Table 5-1, $\delta = 2\pi t/T = \pi/4$ or 45°.
The amplitude is $A = 5\text{ cm} \times 0.1\text{ V/cm} = 0.5\text{ V}$.
The RMS amplitude is $A_{RMS} = A/\sqrt{2} = 0.35\text{ V}$.
The peak-to-peak amplitude is twice the amplitude: $A_{PP} = 2\text{ A} = 1\text{ V}$.

5-4 EXPONENTIAL AND LOGARITHMIC FUNCTIONS

An example of an exponential function in the laboratory is the voltage across a charging or discharging capacitor in an RC circuit:

$$\text{Charging:} \qquad \mathscr{E} - V = \mathscr{E}e^{-t/RC} = \mathscr{E}e^{-t/\tau} \tag{5-13}$$

$$\text{Discharging:} \qquad V = \mathscr{E}e^{-t/RC}$$

where $\tau = RC$ is the time constant of the circuit. Other examples are the growth or decay of current in an RL circuit with time constant L/R and/or the peak voltage swings in a freely oscillating RCL circuit with time constant $2L/R$. Another example is the counting rate of a radioactive isotope:

$$C = C_0 e^{-\lambda t} = C_0 e^{-t/\tau} \tag{5-14}$$

where $\lambda = 1/\tau$ is the *disintegration constant* and τ is the *mean life* of the isotope.

Logarithmic functions occur in electricity in the equipotentials of a pair of concentric cylinders:

$$V(r) = V(a) + \frac{(V_b - V_a)\ln(r/a)}{\ln(b/a)} \tag{5-15}$$

Review of the Mathematics of Exponential and Logarithmic Functions

Law of Exponents.

$$a^P \times a^Q = a^{P+Q}, \qquad (ab)^P = a^P b^P, \qquad (a^P)^Q = a^{PQ} \tag{5-16a,b,c}$$

$$a^0 = 1 \quad \text{if } a \neq 0, \qquad a^{-P} = \frac{1}{a^P}, \qquad \frac{a^P}{a^Q} = a^{P-Q} \tag{5-17a,b,c}$$

$$a^{P/Q} = \sqrt[Q]{a^P}, \qquad a^{1/Q} = \sqrt[Q]{a} \tag{5-18a,b}$$

Law of Logarithms. If M, N, and b are positive numbers and $b \neq 1$,

$$\log_b MN = \log_b M + \log_b N, \qquad \log_b \frac{M}{N} = \log_b M - \log_b N \tag{5-19a,b}$$

$$\log_b M^P = P \times \log_b M, \qquad \log_b \sqrt[Q]{M} = \frac{1}{Q} \times \log_b M \tag{5-20a,b,c}$$

$$\log_b \frac{1}{M} = -\log_b M, \qquad \log_b b = 1, \qquad \log_b 1 = 0 \tag{5-21a,b}$$

Change of base of logarithms ($C \neq 1$).

$$\log_b M = \log_C M \times \log_b C = \frac{\log_C M}{\log_C b} \tag{5-22}$$

The common bases of exponential and logarithmic functions are 10 and

$$e = \lim_{n \to \infty} (1 + 1/n)^n = 2.7182818 \ldots \tag{5-23}$$

Special logarithmic symbols

$$\ln(x) = \log_e x, \qquad \log(x) = \log_{10} x \tag{5-24}$$

$$\ln(x) = \log(x) \times \log_e 10 = 2.3025851 \ldots \log(x) \tag{5-25}$$

$$\log(x) = \ln(x) \times \log_{10} e = 0.4342945 \ldots \ln(x) \tag{5-26}$$

The number 2 enters in the theory of radioactive decay (half-lives):

$$\ln(2) = 0.693\ldots, \qquad \ln\left(\frac{1}{2}\right) = -0.693\ldots,$$

$$\log(2) = 0.301, \log\left(\frac{1}{2}\right) = -0.301 \qquad \text{(5-27a,b,c,d)}$$

Exponentials and logarithms as inverse functions

$$\text{If } y = e^x, \qquad \text{then } x = \ln(y) \tag{5-28}$$

$$\text{If } y = 10^x, \qquad \text{then } x = \log(y) \tag{5-29}$$

Graphs of exponential and logarithmic functions. Figure 5-3 shows the graph of the logarithmic function $\ln(x)$. Note the special points for $x = e = 2.718$ and $x = 1/e = 0.367$ and the various powers of 10.

Figure 5-4 shows the graph of the commonly met decaying exponential function $e^{-t/\tau}$.

5-5 THE DECAYING EXPONENTIAL FUNCTION

The decaying exponential function decays by the same factor f in a time T, *no matter what the starting time*. Thus, in one half-life, the function drops to one-half of its starting value, in two half-lives to one-quarter of its starting value, and so on. *The time interval it takes for an exponential function to change by a certain factor is always the same.*

To put this in mathematical language,

$$\frac{(e^{-(t+T)/\tau})}{(e^{-t/\tau})} = e^{-T/\tau} = \frac{1}{f} \tag{5-30}$$

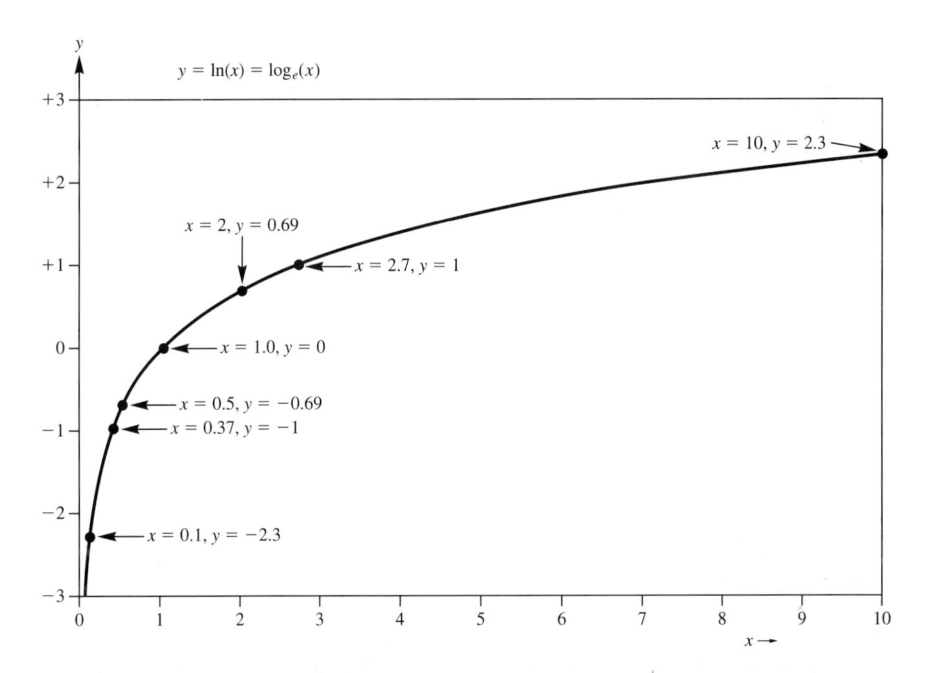

Figure 5-3 The natural logarithmic function $y = \ln(x) = \log_e(x)$. Note the special values of (x, y).

where f is a constant that does not depend on the starting time t. By taking the logarithm to the base e of the last two members of Eq. (5-30), we get a useful expression:

$$\ln(f) = \frac{T}{\tau} \tag{5-31}$$

Some examples of this decay factor are given in Table 5-3. The right side of Fig. 5-5 shows the plot of a decaying exponential function of the form $y = e^{-t/\tau}$. You would get such a curve right from the oscilloscope screen for a discharging RC or RL circuit, for the peak amplitude of a decaying RCL circuit, or from the graph of the counts versus time for a radioactive isotope. The left side of Fig. 5-5 shows the same function inverted, as you might see on the oscilloscope screen for a charging RC or RL circuit or for the growth of radioactivity in an irradiated isotope.

5-6 HOW TO FIND LIFETIMES

A. Quick Methods (Good to ≈ 20%)

1. *Intercept method.* Draw a line TI tangent to your curve at any time T. This line intercepts the time axis of Fig. 5-5 for the right-hand curve (or line $V = \mathscr{E}$ for the curve on the left) at I, one time constant τ later than the time T. You can do this right on the oscilloscope face with a transparent ruler.

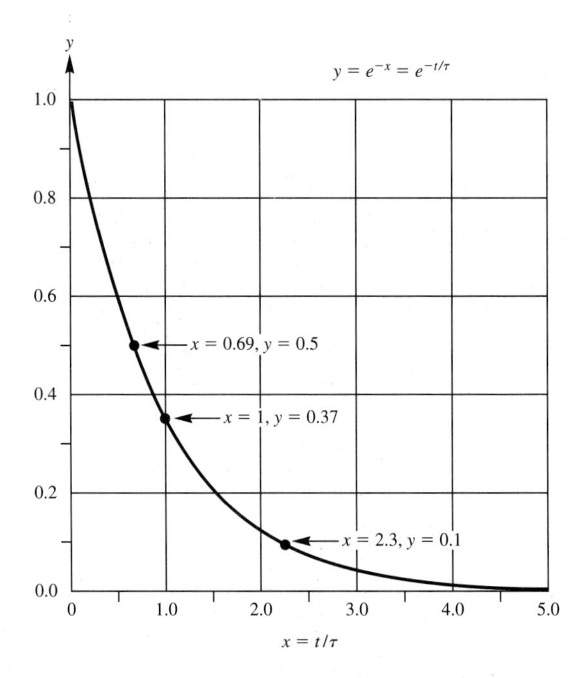

Figure 5-4 The decaying exponential function $y = e^{-x} = e^{-t/\tau}$, where $x = t/\tau$, and τ is the time constant RC or mean life. Note the special points (x, y).

2. *1/e Method.* The time constant is the time that elapses between T and the point labeled $1/e$ on Fig. 5-5. In this interval, $V(\mathscr{E} - V$ on the left) decreases to a value $1/e$ as large ($1/e = 0.367 \approx \frac{3}{8}$).

B. Accurate Methods

More accurate graphical or analytical methods for finding the mean life or time constant usually change the exponential function into a straight line. This makes it much easier to find the best fit of the data points.

Plotting an exponential function as a straight line. Take an exponentially decaying function with mean life (or time constant) τ. Plot its logarithm along the y-axis. This gives you a straight line with equation $y = a + bx$. The slope b of the straight line gives the time constant τ by this relation:

$$b = \frac{-1}{\tau} \tag{5-32}$$

The intercept a is the argument of the exponential function at $t = 0$:

$$e^a = \begin{cases} C_0 \\ V_0 \end{cases} \tag{5-33}$$

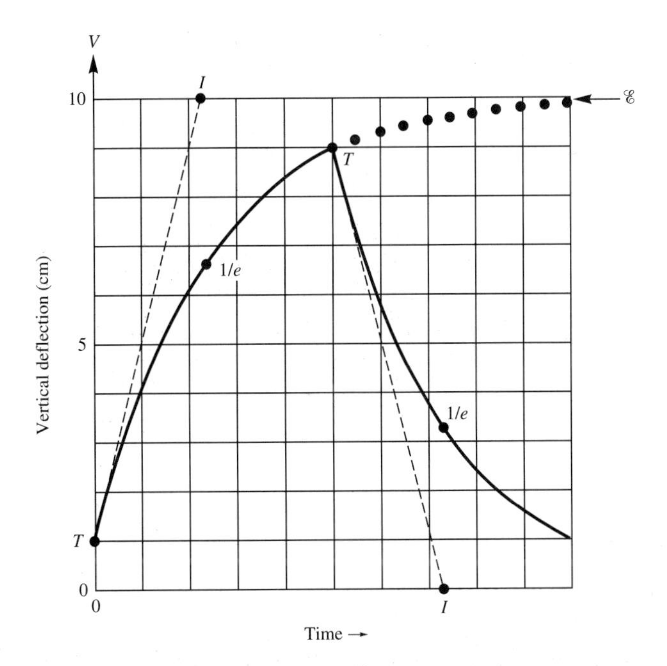

Figure 5-5 Oscilloscope trace of charging (left) and discharging (right) waveform of an RC circuit. The vertical deflection V is proportional to the voltage across the capacitor. The bottom line of the screen is the zero of voltage. The top line of the screen is the asymptote $V = \mathcal{E}$. To find these lines, we must change the conditions of the experiment, such as lengthening the cycle time, so that the dotted extension of the curve shows (as at the upper-right part of the picture). The broken line TI is tangent to the curve at T and reaches the upper or bottom line one time constant later. The points labeled $1/e$ also occur at one time constant after T; the trace at $1/e$ reaches a displacement $1/e = 0.37$ times the displacement at T.

which is the initial number of counts C_0 or initial deflection V_0. If you use semilogarithmic graph paper, you can plot your data directly without taking logs. Also "you see what you get" with less mathematics between you and the answer.

Example 5-5

Plot the natural logarithm ($\ln f$) of the function $f(t) = 10e^{-2.3t}$. Show that the graph is a straight line. Find the intercept and slope. What do they mean? Plot the same data on semilogarithmic graph paper.

Solution. Table 5-2 shows the function and its logarithm, which are plotted in Fig. 5-6 on the left side. The same data are plotted on semilogarithmic graph paper on the right side of Fig. 5-6. Note that, with semilog graph paper, the numbers are plotted directly without finding the logs.

From the graphs (Fig. 5-6), it is evident that the plot is a straight line. We take the left graph and use points 1 and 2 to find the slope $b =$

TABLE 5-2 The Function $10e^{-2.3t} = 10 \times 10^{-t}$.

t	0	0.1	0.2	0.3	0.4	0.5	0.6	0.7	0.8	0.9	1.0
$10e^{-2.3t}$	10	7.9	6.3	5.0	4.0	3.2	2.5	2.0	1.6	1.26	1.0
$\ln_e(10e^{-2.3t})$	2.3	2.07	1.84	1.61	1.39	1.16	0.92	0.69	0.47	0.23	0

$-(y_1 - y_2)/(t_2 - t_1) = -(2.3 - 0)/(1 - 0) = -2.3$. By Eq. (5-32), this gives the mean life $\tau = -1/b = 0.434$. By Eq. (5-33), the intercept $A = 2.3$ gives the value of the function at $t = 0$: $e^{2.3} = 10$.

With semilogarithmic graph paper, it is more convenient to find the mean life by using expression (5-31) directly: $\tau = T/\ln(f)$. We use points 1 and 2 on the right side of the graph, with $f = 10$ and $T = 1$, which give us $\tau = 1/\ln(10) = 0.434$. We read the initial value of $y(0) = 10$ right off the graph.

5-7 DATA ANALYSIS OF EXPONENTIAL FUNCTIONS

A. Using Semilogarithmic Graph Paper

1. *Choose your scales.* Choosing the zero and scale of the x- (or t-) axis is just the same as with ordinary graph paper. But the y-axis is shifty. You can choose the number by which you multiply the scale, such as 1000, 100, 10, 1, 0.1, and 0.01. You can plot your raw data (such as centimeter deflection or number of counts) di-

$f(t) = 10e^{-2.3t} = 10 \times 10^{-t}$

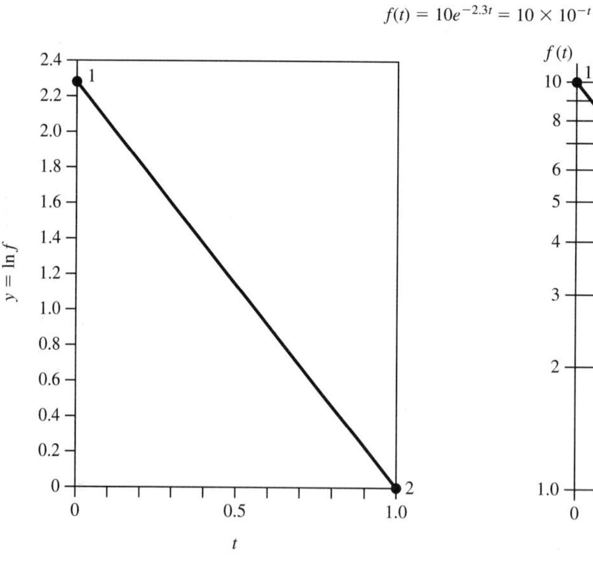

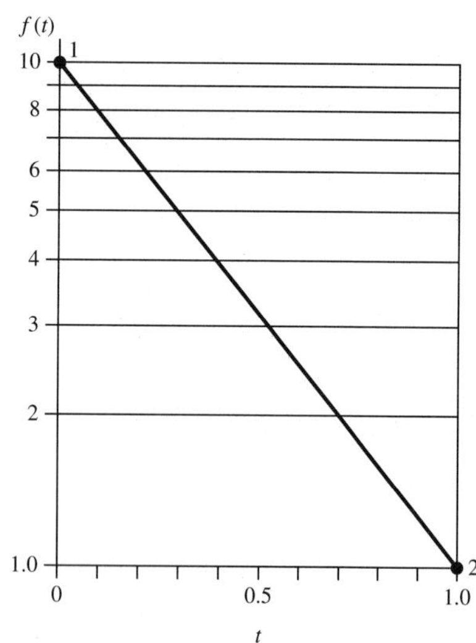

Figure 5-6 The function $f(t) = 10e^{-2.3t} = 10 \times 10^{-t}$ plotted in two, equivalent ways. The inset on the left shows a plot of the natural log on ordinary graph paper. The inset on the right shows the plot of f directly on semilogarithmic graph paper.

rectly on your semilog plot without worrying about units such as volts, millivolts, and counts per second. A change in units, or any other multiplication of y by a constant, merely shifts all the data points up or down by the same amount. Such a displacement doesn't change the shape of a curve or the slope of a line.

How to squeeze or stretch out data on semilog paper. This shifty feature of semilog graph paper sometimes makes it hard to get a good slope for your fitted straight line. What can you do to change the slope if a change of scale only moves the curve up or down? The answer is that you change graph paper. If your curve looks too flat on three-cycle paper, try two- or one-cycle paper. If your data drops through the floor on one-cycle paper, try graph paper with more cycles, or splice together sheets of single cycle paper.

2. *The baseline or background correction for y is important.* If you get a curved semilog plot, your voltage zero or background is likely incorrect. Nevertheless, as Example 5-6 shows, a straight-line semilog plot does not guarantee a correct zero. Always check your background subtraction or oscilloscope zero.

3. *Plot the data.* Figure 5-7 shows such a plot with vertical "error bars" to show the size of the error (both + and −), which is 0.2 cm. Note that the logarithmic plot causes the vertical error bars to become very large for small y (large x). There is then a point of diminishing returns, beyond which it doesn't pay to take more data. Luckily, you've got lots of leeway in choosing this point. Stop when the signal is down to about $\frac{1}{10}$ to $\frac{1}{2}$ of its initial value in oscilloscope experiments, or about 1–25% for counting experiments.

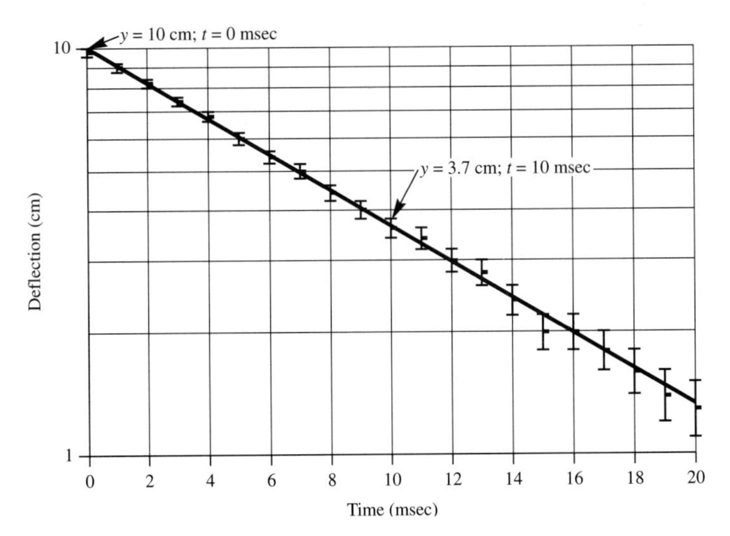

Figure 5-7 Semi-log plot of Nettie's data for the decay of the voltage in an RC circuit.

4. *Fit a straight line to the data points and find the mean life or time constant.* You can follow the method of Section 4-1 to fit a straight line to the set of points and estimate the error in the slope. The intercept of your fitted line with the y-axis gives you the initial count C_0, or the initial deflection V_0 on the 'scope. Take two points on this line (not data points, which are less accurate) and use Eq. (5-31) or Table 5-3 to find the mean life or time constant. Table 5-3 gives some specially useful points. Figure 5-7 shows this procedure for a two points separated by the mean life (y ratio of $1/e = 0.368$).

When using ordinary graph paper, follow the directions of part A, with minor changes. For your y values, plot the natural logarithm of your data and otherwise proceed as in part A.

B. Error Calculations

The fractional error in the time constant is simply equal to the fractional error in the difference quotient b [see expression (5-32)]:

$$\frac{\sigma_\tau}{\tau} = \frac{\sigma_b}{b}, \tag{5-34}$$

or, when the error in the time calibration is taken into account:

$$\frac{\sigma_\tau}{\tau} = \sqrt{\left(\frac{\sigma_b}{b}\right)^2 + \left(\frac{\sigma_t}{t}\right)^2} \tag{5-34'}$$

C. Analytical Methods

Even if you punch out your answer on a keyboard, it still pays to look over your data first with a graph.

1. *Standard Method.* Follow the directions of part A in choosing your data points, especially part A-2 for a correct zero or baseline. Make a table of your data points, with x values equal to time t and y values equal to your measured variable, such as oscilloscope deflection, number of counts or of dividing cells.

If you have a calculator or computer program with built-in exponential regression (see Appendix B), you simply enter your data. If your device lacks ex-

TABLE 5-3 How to Find the Decay Time of an Exponential Function.

Time Interval, T	Decay Factor	Expression for: Mean Life	Expression for: Half-life
General case, T	$1/f$	$T/\ln(f)$	$0.693T/\ln(f)$
Half-life	$1/2$	$1.44T$	T
Mean life	$1/e = 0.368$	T	$0.693T$
Decade	$1/10$	$0.43T$	$0.30T$

ponential regression, you enter x and $\ln_e y$ instead of y, and do a least-squares, linear regression calculation, as in Section 4-2. In either case, you calculate a, b, σ_a, and σ_b. The initial value is given by expression (5-33), the time constant by expression (5-32), the error in time constant by expression (5-34), and the error in the initial count rate C_0 is given by

$$\sigma_{C_0} = \sigma_a C_0 \tag{5-35}$$

for radioactivity experiments. Expression (5-35) holds also for the error in the initial deflection V_0 in oscilloscope experiments.

2. *Weighted Data.* The most accurate (and most complicated) way to handle exponential decay is to weight the data. All data points can be used. You follow the directions for the standard method (given in the previous section), except that each point gets a weight w, which is related to its error. The weight factors are inversely proportional to the square of the errors in y, which are as follows: for oscilloscope data $w = V^2$, where V is the vertical deflection. For counting data, the weight is the number of counts C. See Appendix B for programs that use this method to calculate the lifetime or time constant, the initial count rate or oscilloscope deflection, and the errors in both.

Calculations with weighted data. The expressions for weighted data are

$$b = \frac{\Sigma w(x - \bar{x})(y - \bar{y})}{\Sigma w \sigma_x^2}, \qquad a = \bar{y} - b\bar{x} \tag{5-36}$$

$$\sigma_b = \frac{\sigma}{\sigma_x \sqrt{N - 2}}, \qquad \sigma_a = \sigma_b \sqrt{\frac{\Sigma w x^2}{\Sigma w}}, \tag{5-37}$$

where $\bar{x} = \Sigma wx/\Sigma w$, $\bar{y} = \Sigma wy/\Sigma w$ are the means of the x- and y-values, respectively; σ_x is the standard deviation of the x-values and is given by the expression

$$\sigma_x = \sqrt{\frac{\Sigma w x^2}{\Sigma w} - \bar{x}^2} \tag{5-38}$$

$d = y - (a + bx)$ is the y-deviation of a point (x, y) from the fitted straight line $y = a + bx$, N is the total number of points. Note: $N \neq \Sigma w$), and σ is the y-error, found by the scatter method, for a singly weighted data point; $\sigma = \sqrt{\frac{\Sigma w d^2}{\Sigma w}}$.

Equivalent, but more convenient, formulas are:

$$\sigma_b = b\sqrt{\frac{\frac{1}{r^2} - 1}{(N - 2)}} = b\,\frac{\tan(\cos^{-1} r)}{\sqrt{N - 2}} \tag{5-39}$$

In the case of fitting an exponential function to a decaying or growing number of counts, one can estimate in advance the error in b by the expression

$$\sigma_b = \frac{1}{\sigma_x \sqrt{\Sigma w}} \qquad (5\text{-}37')$$

where Σw is the total number of counts in the entire experiment.

5-8 MEASURING THE TIME CONSTANT OF AN *RC* CIRCUIT

A square-wave generator alternately charges and discharges a capacitor in a *RC* circuit. The waveform on an oscilloscope is shown in Fig. 5-5. You can find the time constant of the circuit from either the discharge or charge waveforms by using Eq. (5-13).

You analyze either set of data the same way. Both parts of Eq. (5-13) lead to the same form of data, exponentially decaying functions with the same time constant *RC*.

To find the baseline, set the period of the square wave long enough so that the curve can level off before the next part of the cycle (see Fig. 5-5). The (dis)charge time must be at least 5 time constants.

A second method is to change the circuit to show zero voltage and the value $\mathscr{E}$ right on the screen. (You can do this, for example, by removing the capacitor or by shorting out the charging resistor *R*.)

Set the zero voltage right on the bottom line of the oscilloscope screen. Then you can measure the voltage directly off the screen without tiresome and erroneous subtractions. (Likewise, on charging measurements, set $V = \mathscr{E}$ right on the top line and measure $\mathscr{E} - V$ directly off the screen by measuring *down* from the top line.)

A further remark. Remember that the measured deflection y is the *difference* between the y value of the trace and the y value of the zero voltage for decay, or the y value of the asymptote in charging. Any error in setting your zero (or your asymptote) will add to your error in y. Be careful!

Example 5-6 Measurement of the Time Constant of an *RC* Circuit

Nettie Neat uses an oscilloscope to measure the decay of the capacitor voltage in an *RC* circuit. Figure 5-7 shows a semilog plot of her initial data. Nettie decides the points beyond 10 msec are too noisy (see Fig. 5-7) and discards them. She starts at $y = 10$ cm, $t = 0$ and ends at $y = 3.7$ cm, $t = 10$ msec. She measures 11 y values to the nearest 2 mm, with an x calibration of 1 msec/cm (1% error). She uses an Excel™ program, *FitExponential* to analyze her data. The data, fits, and semilog plot are shown in Fig. 5-8. The program finds the time constant and initial deflection and estimates the errors due to scatter. She also finds the error in the time constant *RC*, which has 1% due to data scatter in addition to the 1% error in the time axis. Her final error in *RC* is the root-square sum, or 1.4%. Appendix C gives a table with all the details of the calculation.

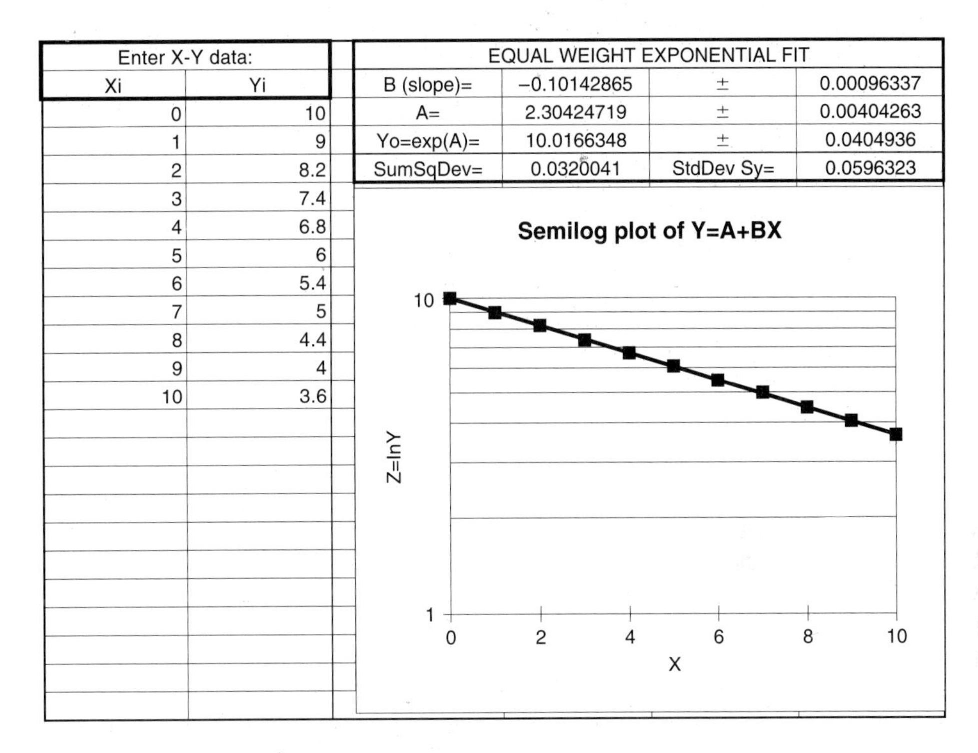

Enter X-Y data:	
Xi	Yi
0	10
1	9
2	8.2
3	7.4
4	6.8
5	6
6	5.4
7	5
8	4.4
9	4
10	3.6

EQUAL WEIGHT EXPONENTIAL FIT			
B (slope)=	−0.10142865	±	0.00096337
A=	2.30424719	±	0.00404263
Yo=exp(A)=	10.0166348	±	0.0404936
SumSqDev=	0.0320041	StdDev Sy=	0.0596323

Figure 5-8 A fit to Nettie's Oscilloscope data.

Nettie's partner Lassy Lazy does the same experiment, but is less careful than Nettie in zeroing her 'scope. Figure 5-9 shows the 'scope traces for the two lab partners for times from zero to 10 msec. Both traces have exactly the same shape and look equally good to the eye. Figure 5-10 shows Lassy's data, fits and the semilog graph. Again, there is nothing in the graph that shows a difference in the quality of the two fits. Lassy's error in b is only slightly larger, 2% versus 1.0%, hardly cause for suspicion. Yet the two values differ by almost 20%! The percent standard error of the difference is $\sqrt{(1^2 + 2^2)} = 2\%$. The difference is ten times as large as its error, a highly significant result.

None of the graphs or fits shows a difference in the data quality, yet the disagreement in b is real. Nettie's and Lassy's lab instructor suggests that they look at the data for longer times. They turn the time axis knob two steps, for a longer sweep of 50 msec, which is five time constants. The result is shown in Figure 5-11, which reveals the cause of the discrepancy. Lassy's scope was improperly zeroed, which caused a systematic error that neither graphs nor statistical fits revealed. Lassy's measurement of the time constant RC was no improvement over just reading the labeled values, which are nominally good to 20% for C and 10% for R, which yield the same error in the time constant RC of 20%.

The moral of this story is that you cannot trust a statistical analysis to catch a poorly done experiment. Graphs often will reveal systematic errors missed by the fitting procedures, but are not infallible.

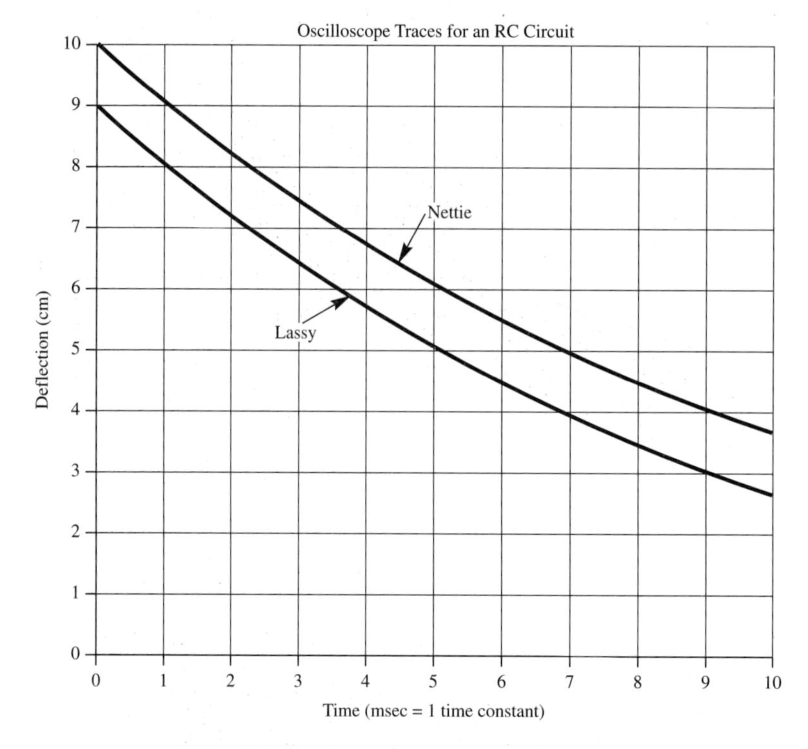

Figure 5-9 Nettie's and Lassy's Oscilloscope data.

5-9 MEASURING RADIOACTIVE ISOTOPE LIFETIMES. POISSON DISTRIBUTION

As discussed in the introduction to this chapter, the counting rate of a radioactive isotope is an exponentially decaying function of time. To find this rate, you repeatedly measure the number of counts in a radiation detector for a certain period of time and subtract the background counts from your data to get the true number of counts C. With this as data, you can use the graphical (see Section 5-A,B) or analytical method (see Section 5-C) to find the initial counting rate C_0, the decay (time) constant, and the half-life (see Table 5-3).

Errors. Poisson Distribution

Imagine that it started to rain and you walked down the sidewalk and counted the number of raindrops that hit each square (see Fig. 5-12). If 6 drops hit one square, would the number be exactly 6 in the next square? If not, how much would it differ from 6?

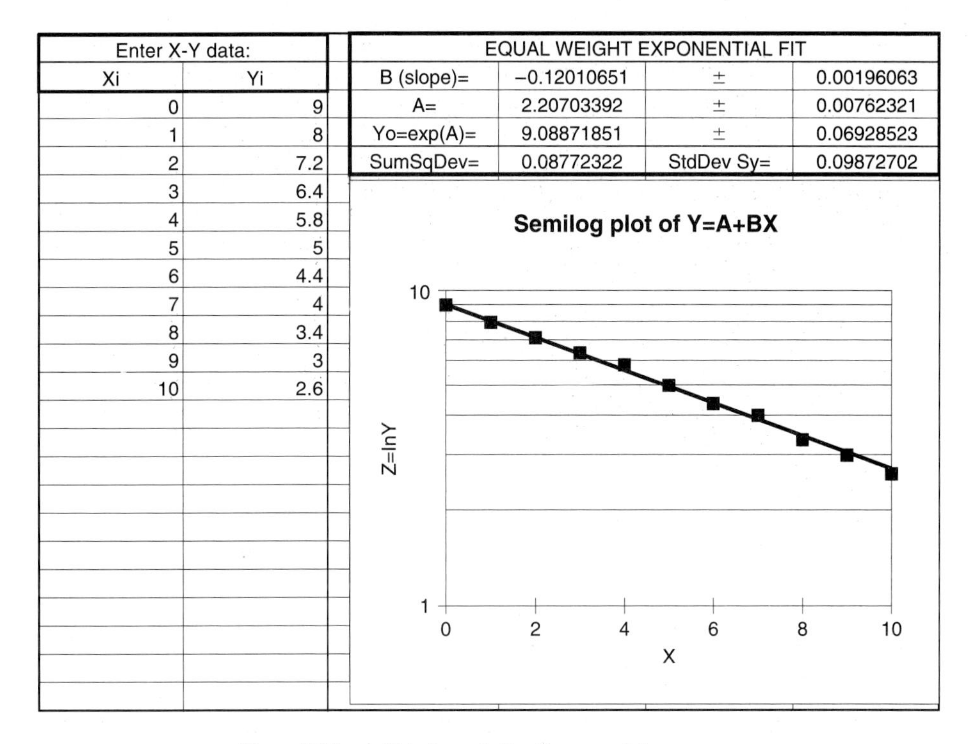

Enter X-Y data:	
Xi	Yi
0	9
1	8
2	7.2
3	6.4
4	5.8
5	5
6	4.4
7	4
8	3.4
9	3
10	2.6

EQUAL WEIGHT EXPONENTIAL FIT			
B (slope)=	−0.12010651	±	0.00196063
A=	2.20703392	±	0.00762321
Yo=exp(A)=	9.08871851	±	0.06928523
SumSqDev=	0.08772322	StdDev Sy=	0.09872702

Figure 5-10 A fit to Lassy's Oscilloscope data.

Just as in public opinion polls, a count of random events is not exact, but is a matter of probability. If one square had 6 drops, you might expect to find 5 or 7 drops in the next; but were you to find 100, you would be surprised. The Poisson distribution tells you more exactly: if C is the average number of counts, the probability P of finding x counts is given by

$$P = C^x e^{-c}/x! \tag{5-40}$$

The standard deviation of this distribution gives the error in C counts:

$$\sigma_c = \sqrt{C} \tag{5-41}$$

Figure 5-13 shows the distribution of raindrops on the sidewalk (Poisson distribution for $C = 6$).

In a radioactivity experiment, expression (5-41) gives the counting error, if the error in the background is negligible. The error bars for sample data are shown in Fig. 5-14. Thus the weight factor w, inversely proportional to the square of the error in $y = \ln(C)$, is equal to the number of counts C.

With this substitution, expressions (5-37) and (5-38) give the decay constant $b = 1/t = 0.693/t_{\frac{1}{2}}$ and its error by the scatter method. An *a priori* (in advance) formula gives an alternative error estimate, which does not depend on

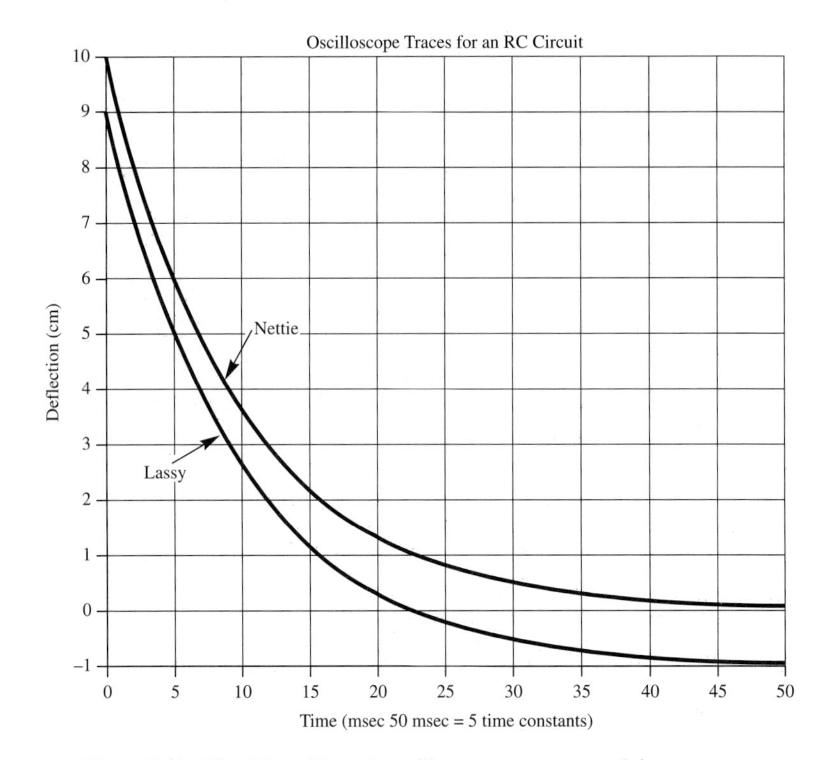

Figure 5-11 Nettie's and Lassy's oscilloscope traces out to 5 time constants.

data scatter. It is

$$\sigma_b = \frac{1}{\sigma_x \sqrt{\Sigma w}} \tag{5-37'}$$

Most fitting programs use the scatter method.

Example 5-7 Finding the Mean Life of a Radioactive Isotope

Zeke Zahler measures the number of counts for a radioactive isotope over several 10-minute periods. He subtracts the background counting rate from each set of results to get the true rate C. Zeke uses the Excel™ program *FitCount*. His data, fits and a semilog plot of both are shown in Figure 5-14. Appendix C shows the details of the computations. The results are $C_0 = 1020(30)$ and $\tau = 19.6(5)$ min.

5-10 PLOTTING AND CALCULATING POWER LAW RELATIONS

Often one has two variables connected by a power law relationship

$$y = Ax^B \tag{5-42}$$

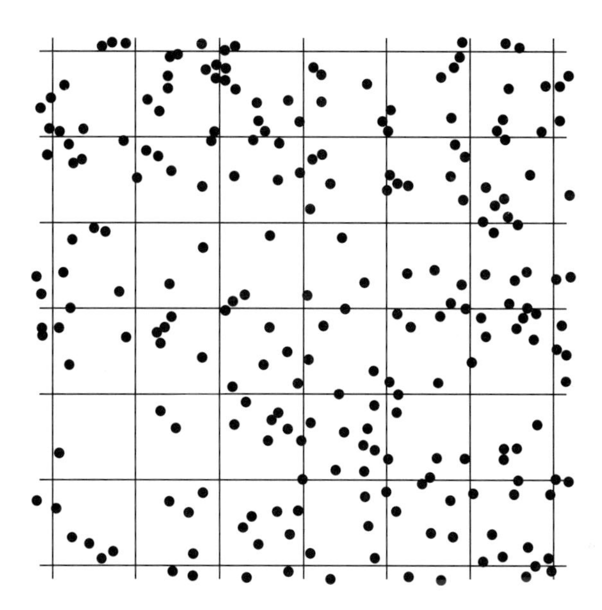

Figure 5-12 A random pattern of dots. The average number in each square is $C = 6$. The number of dots per square follows the Poisson distribution. This is the type of pattern formed by raindrops on the sidewalk or blood cells on a microscope slide.

We consider some examples from physics and chemistry. First, given the period T of a simple pendulum, find the length L. From the formula for the period of a pendulum:

$$T = 2\pi\sqrt{\frac{L}{g}},$$

we have

$$L = g\left(\frac{T}{2\pi}\right)^2,$$

with $A = g/(2\pi)^2$ and $B = 2$. Figure 5-15 graphs this relation.

Second, given the volume V in liters of a Florence (spherical) flask, find the diameter d. From the expression for the volume of a sphere (we neglect the volume of the walls and of the neck)

$$V = \frac{\pi}{6}d^3,$$

which gives us

$$d = \sqrt[3]{\frac{6V}{\pi}},$$

with $A = \sqrt[3]{\frac{6}{\pi}}$ and $B = \frac{1}{3}$. This relation is graphed in Figure 5-16.

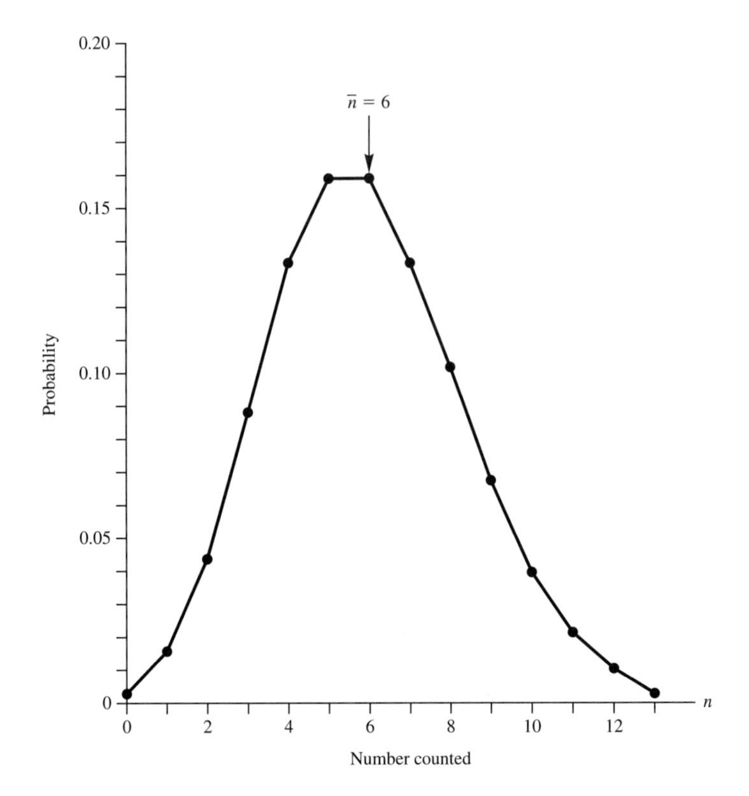

Figure 5-13 The chances of counting N raindrops on a square of the sidewalk, where the average numbers of drops counted is $C = 6$. This is the Poisson probability distribution, which holds for any random counting process, such as measuring the lifetime of a radioactive isotope or making a blood count.

As a last example, we measure the mass M of a steel ball of diameter d. The mass is given by the expression $M = \rho V = \frac{\pi}{6}\rho d^3$, with $A = \pi\rho/6$ and $B = 3$. The data are graphed in Figure 5-17.

The plot of a power law relation $y = Ax^B$ on log-log graph paper gives a straight line of slope $\Delta y/\Delta x = B$. Figure 5-18 shows the log-log plots of the three examples of Figures 5-15, 5-16, and 5-17.

Figure 5-19 determines the slope for the log mass of the steel ball versus the log diameter (Fig. 5-17). With a ruler, one measures the ratio $\Delta y/\Delta x = 3$. [Note: Use commercial log-log graph paper where one cycle (factor of 10) has the same length for both x- and y-axes. It is often difficult to make computer log-log plots with equal x- and y-scales.] The log mass-log diameter relation for the ball has a slope of 3, since the mass of a ball is proportional to the cube of its diameter, i.e., to its volume. The slopes for the pendulum (Fig. 5-15) and for the flask (Fig. 5-16) have the expected values of 2 and $\frac{1}{3}$, respectively.

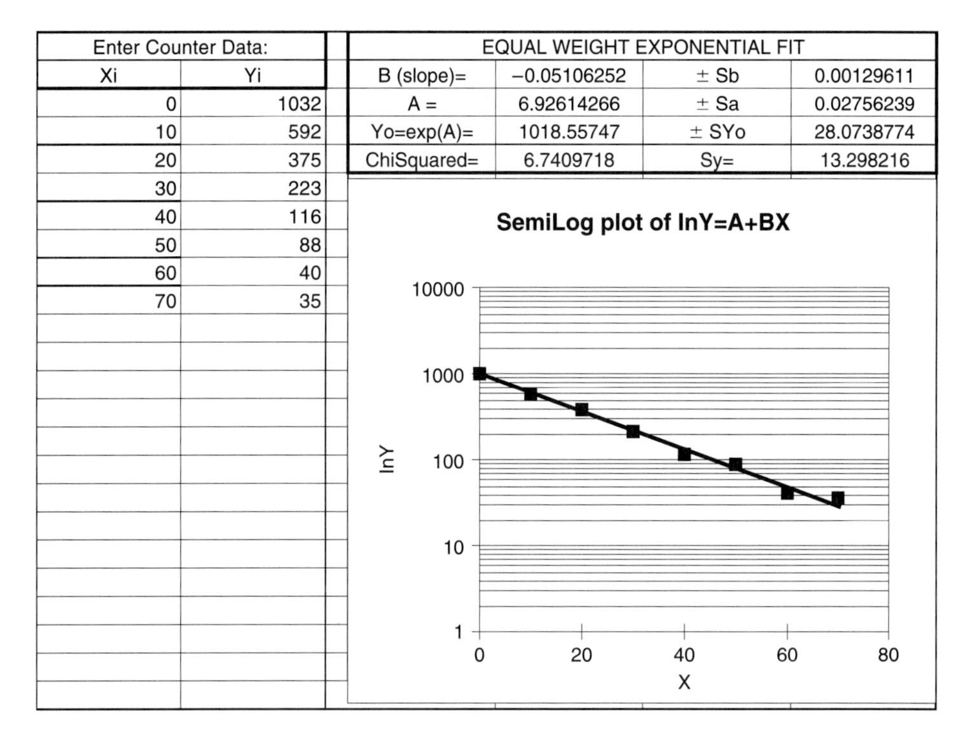

Enter Counter Data:	
Xi	Yi
0	1032
10	592
20	375
30	223
40	116
50	88
60	40
70	35

EQUAL WEIGHT EXPONENTIAL FIT			
B (slope)=	−0.05106252	± Sb	0.00129611
A =	6.92614266	± Sa	0.02756239
Yo=exp(A)=	1018.55747	± SYo	28.0738774
ChiSquared=	6.7409718	Sy=	13.298216

Figure 5-14 Zeke's counter data fit with an exponential function.

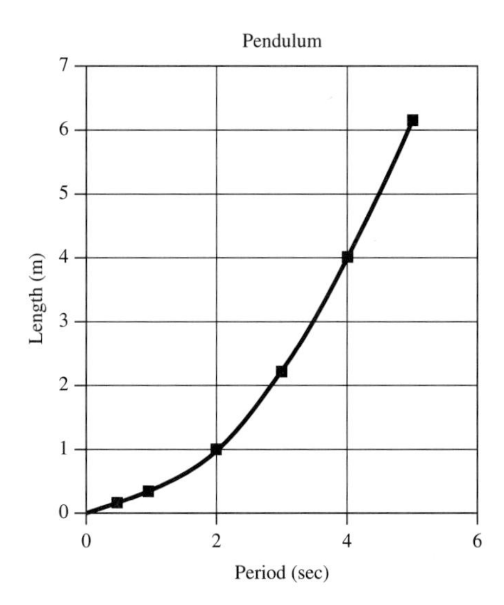

Figure 5-15 The length of a pendulum is proportional to the square of its period.

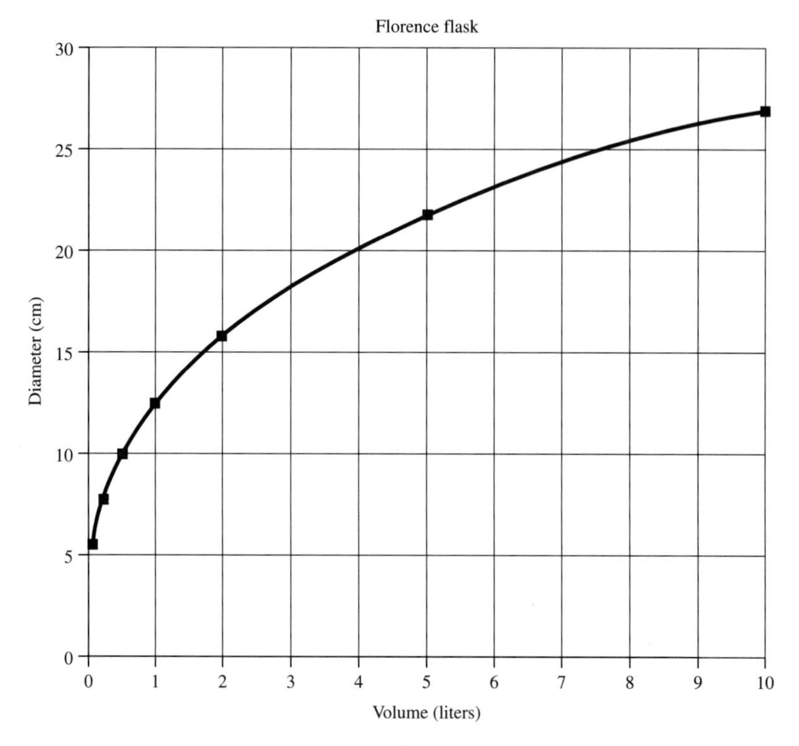

Figure 5-16 The diameter of a Florence flask is proportional to the cube root of its volume.

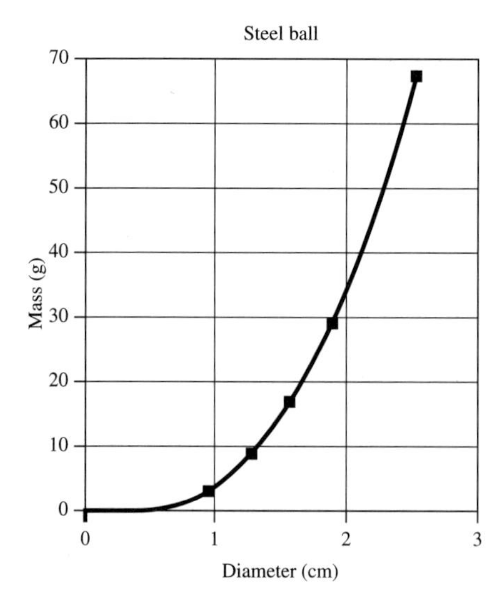

Figure 5-17 Mass and diameter of a steel ball.

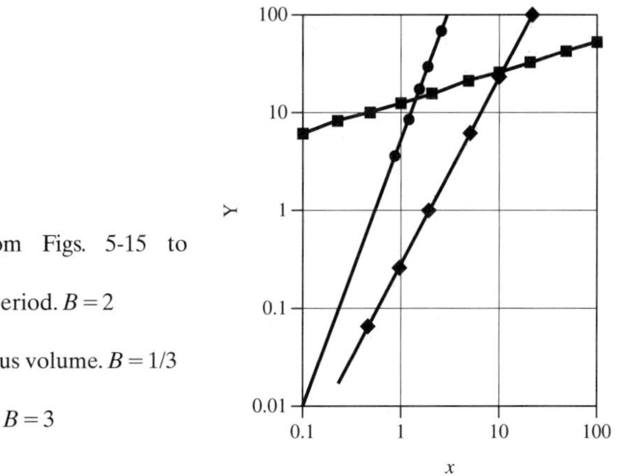

Figure 5-18 Plots of data from Figs. 5-15 to 5-17 on a log-log graph paper.
◆----◆ length of pendulum versus period. $B = 2$ (Fig. 5-15).
■~~~~~~~■ diameter of flask versus volume. $B = 1/3$ (Fig. 5-16).
•-----• mass of ball versus diameter. $B = 3$ (Fig. 5-17).

Other methods of finding a power law. If you do not have log-log graph paper, get $\ln(y)$ and $\ln(x)$ with your calculator and plot them on ordinary graph paper. You will get the same graph as Fig. 5-19, from which you can measure the slope. You can also put these values in a linear regression program in your calculator or computer and solve for the slope B, which is the power in Eq. (1). Many calculators, which have built-in programs for finding the power B in Eq. (5-42), use the algorithm of fitting a straight line on a log-log plot.

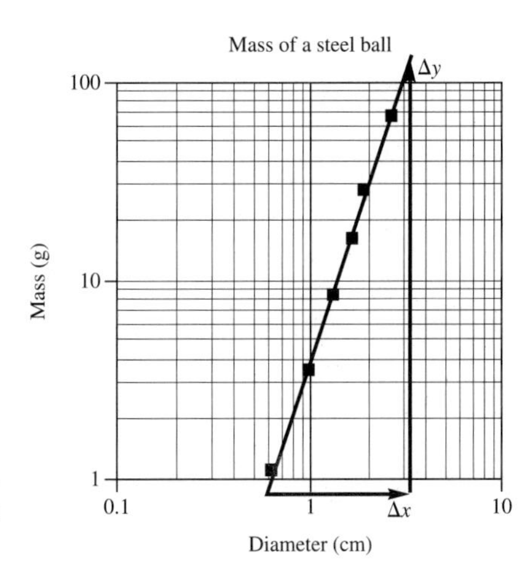

Figure 5-19 Finding the slope B of a log-log plot. B is equal to the power of x in Eq. (1).

A more accurate method is to weight the points, with larger weight for larger y. The Excel programs *FitMac.XL5, FitPc.XL5, FitPwrLaw.XL4,* and *FitPwr.XL4* weight points proportionally to y^2, which is the correct way of fitting a power law relation.

Example 5-8 Jack Measures the Current-Voltage Curve for a Lightbulb

Jack and Jill used a flashlight bulb to study several laws of physics (see Problem 4-12). Jack measured the current-voltage relation and made a log-log plot, with the hypothesis that it would fit a simple power law. To his disappointment, the graph disagreed with his hypothesis (Fig. 5-20). He then looked for straight sections of his curve and used a power-law fitting program, such as *PwrLawFit* (see Appendix B for other programs). The first five points fit a straight line with slope $B = 1.11 \pm 0.02$ and the last five points fit a line with $B = 1.83 \pm 0.003$.

The lower line occurs for low currents, where Ohm's law $V = IR$ would be expected to give a straight line relation, with a slope of $B = 1$ on the log-log plot. The measured value is close to that value, although a small difference arises from the variation of resistance R with current.

The upper line occurs when the bulb is so hot that it is glowing. In such a case, the power goes almost completely into radiation. From the Stefan-Boltzmann radiation relation ($P \alpha T^4$) and the proportionality of resistance to temperature (see Jill's measurements in Problem 4-12), one would expect a $\frac{5}{3}$ power law. Although the measured slope is close to that value, it differs slightly from the theoretical value, due to oversimplified assumptions in the derivation of the $\frac{5}{3}$ power law.

This is an example of where a graph, rather than a straight computer fit, reveals the not-so-simple physics of an ordinary object, a flashlight bulb.

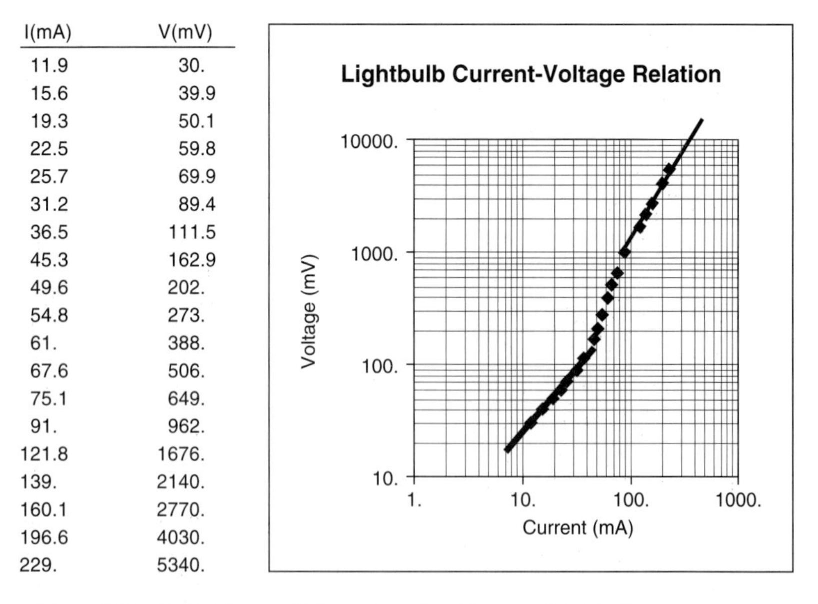

I(mA)	V(mV)
11.9	30.
15.6	39.9
19.3	50.1
22.5	59.8
25.7	69.9
31.2	89.4
36.5	111.5
45.3	162.9
49.6	202.
54.8	273.
61.	388.
67.6	506.
75.1	649.
91.	962.
121.8	1676.
139.	2140.
160.1	2770.
196.6	4030.
229.	5340.

Figure 5-20 Jack's Current-Voltage Data for a Tungsten Lightbulb.

5-11 A COMMENT ON METHOD

If you try more than one method for calculating the error of the *RC* time constant or radioactive lifetime, you may find your results do not agree. This does not necessarily mean you are mistaken. In particular, the weighted regression method (counts) *assumes* a certain error; the standard method of Chapter 4 uses the *actual* scatter of the data from a straight line. This is another example of the inaccuracy of error estimates!

Summary

Units and conversion factors for angles are:

Radians-degrees: 1 rad $= (180/\pi)^\circ = 57.29578^\circ$

Degrees-minutes-seconds($^\circ$, $'$, $''$): $1^\circ = 60'$; $1' = 60''$

Sinusoidal waveforms: Let $y = A \sin(\omega t - \delta)$, with A = amplitude, ω = angular velocity $= 2\pi f = 2\pi/T$, f = frequency $= 1/T$, T = period, δ = phase shift

Amplitude: Root-mean-square (ac) $A_{RMS} = A/\sqrt{2} = 0.707$ A
Peak-to-peak $= A_{PP} = 2$ A

Exponential Functions

RC circuit: $V = \mathscr{E}e^{-t/RC}$; $\ln V = \ln \mathscr{E} - t/RC$

Radioactive decay: $C = C_0e^{-\lambda t} = C_0e^{-t/\tau}$; $\ln C = \ln C_0 - t/\tau$

Half-life $\tau_{1/2} = 0.693\,\tau$

Measurement of Mean Life or Time Constant

1. *Quick:* Find $1/e$ point, tangent intercept.
2. *Accurate:*
   a. Graphical: Use semilog graph paper, or plot ln (V or C) on linear graph paper versus time. Use simple ratios to find lifetime.
   b. Analytical: Use least-squares fit of points to linear relation of ln V or ln C versus time. Weighted points give the best fit.

Errors: $\frac{\sigma_\tau}{\tau} = \frac{\sigma_b}{b}$, $\sigma_{C_0} = \sigma_a C_0$ (likewise for σ_{v_0}), $\sigma_b = \frac{\sigma}{\sigma_x\sqrt{N-2}}$, $\sigma_a = \sigma_b\sqrt{\frac{\Sigma wx^2}{\Sigma w}}$, $\sigma = \sqrt{\frac{\Sigma wd^2}{\Sigma w}}$ (scatter method) $\sigma_b = 1/\sigma_x\sqrt{\Sigma w}$, where weight w = C for radioactivity (*a priori* method)

Baseline subtraction. Subtraction of oscilloscope zero from data, or data from asymptote, and radioactive background subtraction must be made to avoid systematic errors.

Problems

5-1. A wheel makes exactly one rotation. (a) By how many degrees does it turn? (b) By how many radians (in decimal form)?

5-2. In a measurement of the angular separation between two orders, a spectrometer rotates from 159° 12′ to 161° 8′. (a) What is the deflection θ (difference between the two angles) in degrees and minutes? (b) In decimal degrees? (c) In radians? (d) What is $\sin(\theta)$?

5-3. Use your calculator to find the angle that satisfies the equation (a) $\sin(\theta) = 0.6$, (b) $\cos(\theta) = 0.6$, (c) $\tan(\theta) = 0.6$.

5-4. Find the answer to "How many x are there in y?" for (a) seconds in a day, (b) hours in 10,000 seconds, (c) seconds of arc in a degree, (d) minutes in a degree, (e) decimal degrees in 40′, (f) radians in 10°, (g) degrees in 0.02 radian.

5-5. For the sinusoidal wave in Fig. 5-21, find the (a) peak-to-peak amplitude (in volts), (b) amplitude (in volts), (c) RMS amplitude (in volts), (d) period (in seconds), (e) frequency (in hertz = $\sec^{-1}$), (f) angular frequency in radians per second.

***5-6.** Use the power series [expressions (5-4), (5-5), and (5-6)] up to and including the first term involving θ to find the sine, cosine, and tangent of $\theta = 10°$. (b) Use your calculator to find the correct value. (c) Find the percentage deviation between parts (a) and (b) in each case. (d) Use the next term in the series to estimate the percentage error and compare your results with part (c).

5-7. Find the value of (a) $\ln(2)$, (b) $\log(2)$, (c) $\ln(\frac{1}{2})$, (d) $\log(\frac{1}{2})$, (e) $10^{0.5}$, (f) e^{-2}, (g) $e^{0.2}$.

5-8. Use the graph in Fig. 5-6 and Table 5-3 to find (a) the time it takes 10^{-t} to decay by a factor of 10, (b) to decay by a factor of $e = 2.7128\ldots$, (c) to decay by a fac-

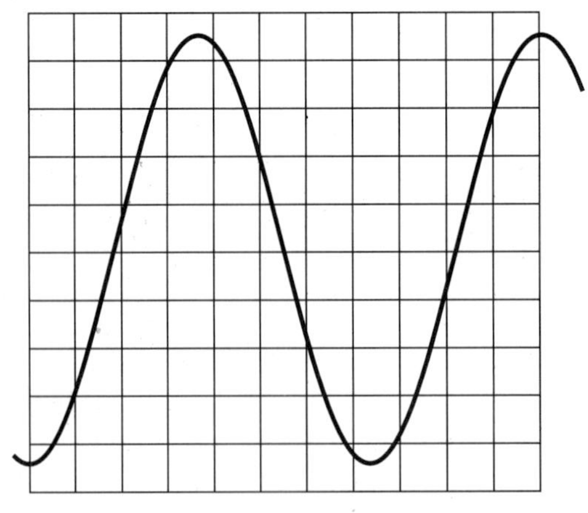

Figure 5-21

tor of 2, (d) the amount the function decays in $\frac{1}{2}$ sec ($\Delta t = \frac{1}{2}$). Assume one unit of t equals 1 sec. Choose a pair of points, give the t and y values for each point, and show your calculations.

5-9. Do the same as for Problems 5-8, parts (a), (b), and (c), from the line fitted to Zeke's data points in Fig. 5-14. For part (d), take a time interval of 30 min.

***5-10.** A student uses an oscilloscope to measure the voltage across a discharging capacitor. His data are shown in Table 5-4. Plot his data on semilogarithmic graph paper (or plot the logarithm of voltage on the y-axis of ordinary graph paper) and test to see if it fits a straight line. If it does not, try to see if you can correct his data by changing the zero of the y-axis, by subtracting or adding a fixed amount to all his y values. Use your corrected data to find the time constant (and error) for the circuit. Assume a least count of 0.1 cm.

TABLE 5-4 Oscilloscope Data for Voltage Decay in an *RC* Circuit.

x	0	1	2	3	4	5	6	7	x: 1 division = 20 μsec
y	8.5	5.8	3.7	2.6	1.75	1.3	1.0	0.8	y: 1 division = 0.1 V

***5-11.** A medical technician takes a blood count by diluting a drop and spreading it on a special microscope slide (see Fig. 5-12), which is divided into compartments. Each compartment holds 1/800,000 mm^3 of undiluted blood. The technician counts six blood cells in one compartment. (a) How many blood cells are in a cubic millimeter of whole blood? (b) What is the error in this estimate? (c) What would be the error if the technician counted 600 blood cells in 100 compartments? (d) Calculate the probability distribution of finding m cells in a given slide compartment.

***5-12.** The vapor pressure of a liquid obeys the equation $\ln(p) = a - b/T$, where a is a constant and $b = \Delta H/R$, where ΔH is the heat of vaporization in cal/mole K, $T = 273.15 + T(°\text{C})$ is the absolute temperature, and $R = 1.99$ cal/mole K is the universal gas constant. Table 5-5 gives the vapor pressure of water as a function of T(°C). Find the heat of vaporization of water. (Hint: fit a straight line of $\ln(p)$ versus $1/T$ and find b, or fit p to an exponential function of $1/T$.)

TABLE 5-5 Vapor Pressure of Water as a Function of Temperature.

T(°C)	0	10	20	30	40	50
p(mm Hg)	4.579	9.209	17.535	31.824	55.324	92.51

***5-13.** A public health laboratory studies a new bacterium by injecting it into a test site, and taking periodic samples from the test site. Table 5-6 shows the growth of number of cultures counted as a function of time. Find the division time for the bacteria.

TABLE 5-6 Colony Count of a Culture of Bacteria.

Time (hr)	0	1	2	3	4	5	6	7	8	9
Number	1	2	9	14	50	135	280	521	748	913
Time (hr)	10	11	12	13						
Number	956	1023	982	1014						

FURTHER READING

Brief

H. MARGENAU and G. M. MURPHY, *The Mathematics of Physics and Chemistry* (D. Van Nostrand, New York, 1943; 2d ed., 1956). Chapter 13, Part 3 (Sections 13.29–13.37) has a concise summary of error theory, especially a proof of expression (2-2′).

General Reference

W. SNEDECOR and W. G. COCHRAN, *Statistical Methods*, 8th ed. (Iowa State University Press, 1989).

Detailed

P. R. BEVINGTON and D. K. ROBINSON, *Data Reduction and Error Analysis for the Physical Sciences*, 2d ed., (McGraw-Hill, New York, 1992).

Present Book	Bevington and Robinson
Chapter 1	Chapter 1
Chapter 2	Chapters 2, 4
Chapter 3	Chapter 3
Chapter 4	Chapter 6
Chapter 5	Section 7-4; Sections 2-2, 5-4 (Poisson distribution)

Appendix A:

SEMI-LOGARITHMIC AND LOGARITHMIC GRAPH PAPER

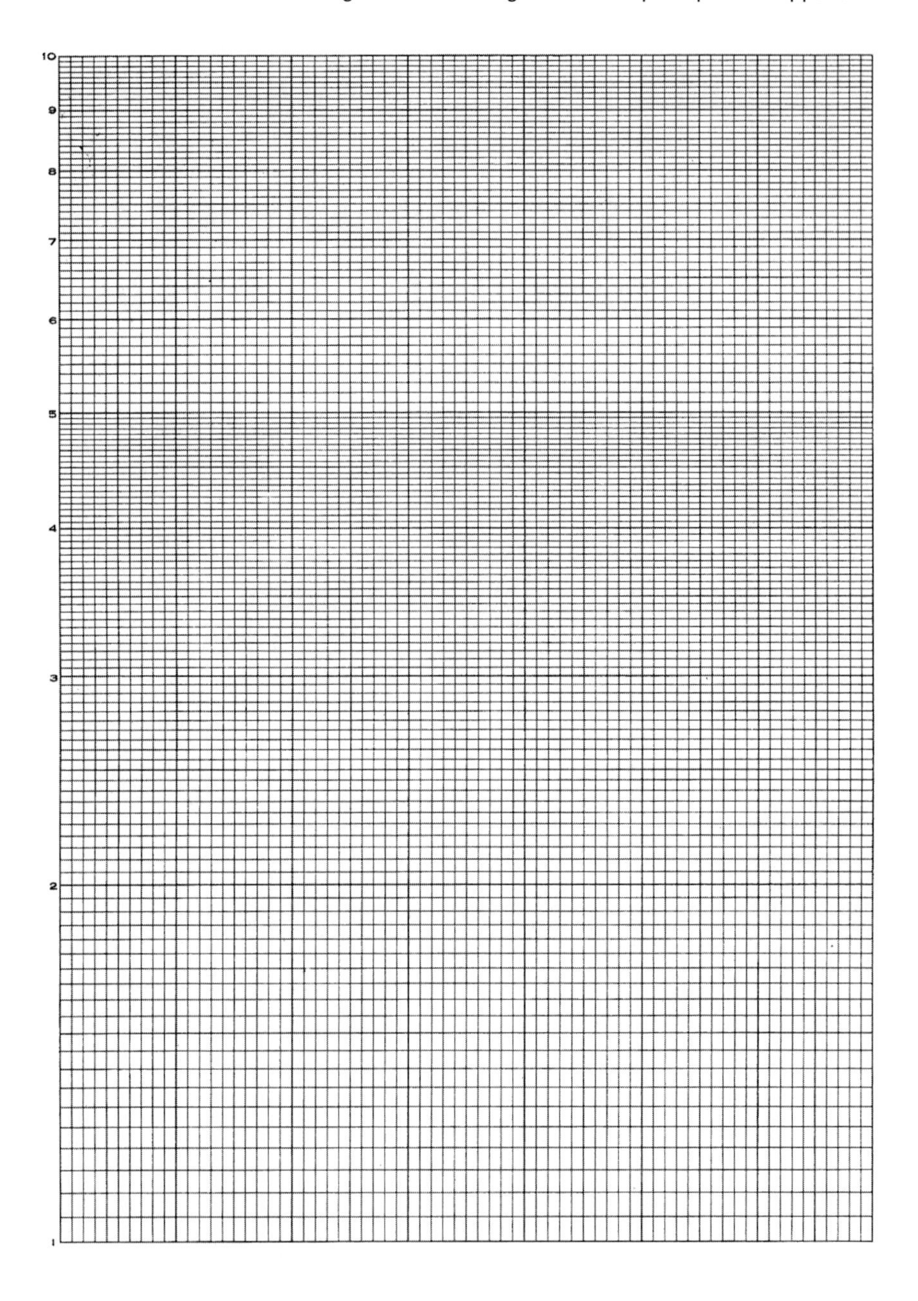
10
9
8
7
6
5
4
3
2
1

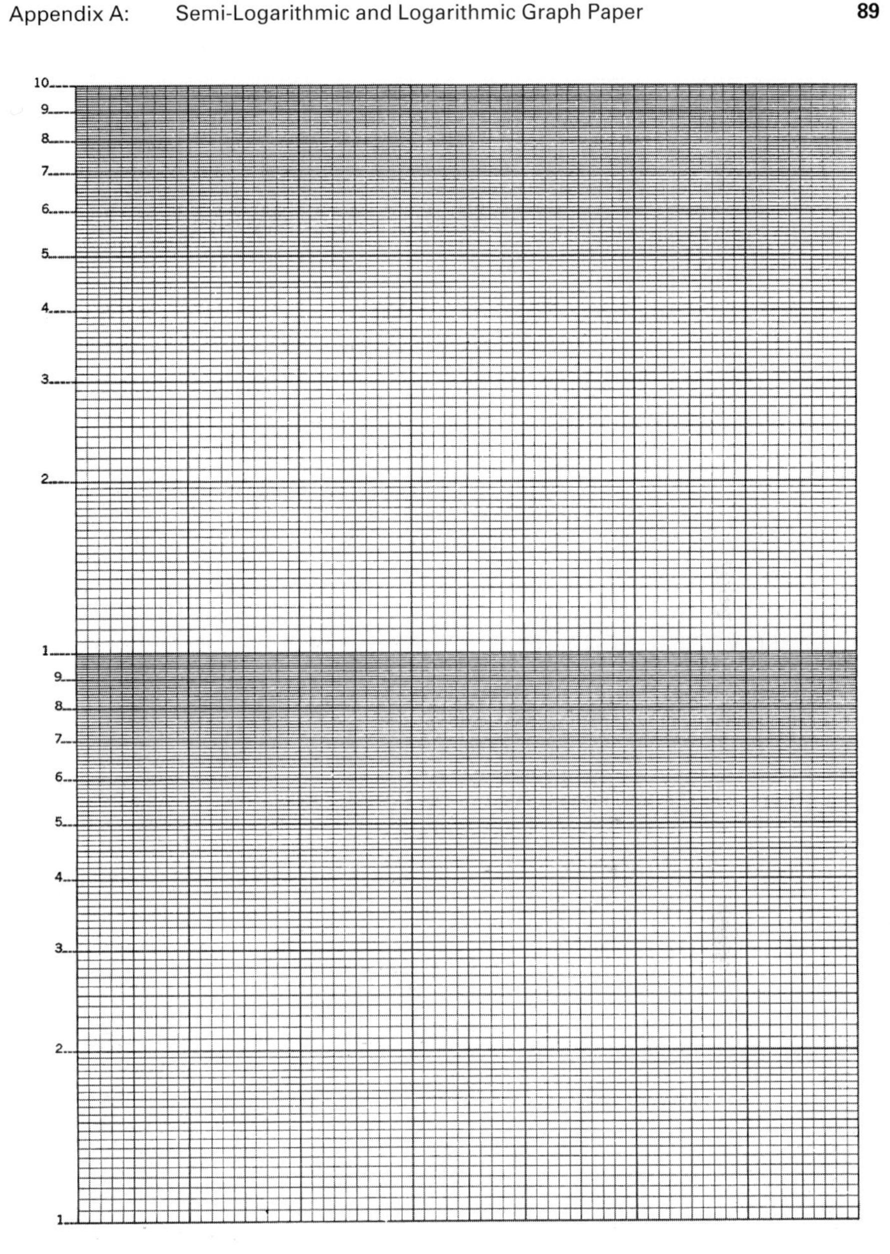
10
9
8
7
6
5
4
3
2
1
9
8
7
6
5
4
3
2
1

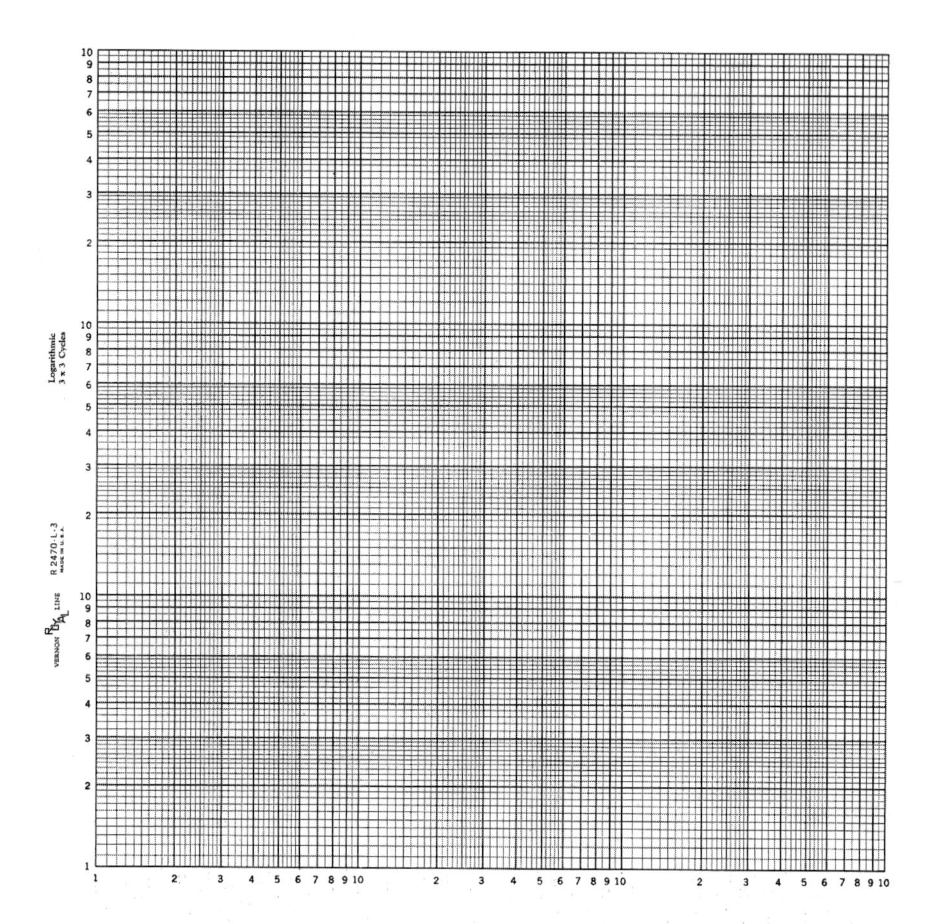
Logarithmic
3 x 3 Cycles
R 2470-L-3
MADE IN U.S.A.
VERNON LINE

Appendix B:

CALCULATOR AND COMPUTER PROGRAMS

CALCULATOR		PROGRAM OR SET OF INSTRUCTIONS				
MODEL	TYPE	St. Dev. Grouped data	Linear Regr., errors	Trig Con-versions	Log, Exp, Pwr Regr.	
					Unwgted	Weighted
Reference→	Section→	2-6	4-2	5-1,2	5-7, -10	5-7, -10
Casio fx-100S, 115S 570S, 991S 300SA, 85SA	Scientific	I	I	I	I	I
fx-7400G, fx-9850G.	Graphics	I	I	I	I	I
Hewlett Packard 20S	Program-mable	I	I	I	I	
HP-48G	Graphics	I	I	I	I	
Sharp EL-506L, 520L, 546L	Scientific	I	I	I	I	I
TI-80, 82, 83, 86	Graphics	I	I	I	I	I
TI-81, 85	Graphics	I	I	I	I	
Basic, Pascal	Computer Languages	I	I	I	I	I
Excel™	Computer Spread-sheet	I	I	I	I	I

Table A-1. Programs or instructions included in this book (marked "I").

Explanations: All listed calculators have built-in programs to do unweighted standard deviations and linear regression.

Scientific calculators are not programmable; the sequence of buttons must be pushed for each calculation.

Programmable calculators remember the sequence of buttons to be pushed and can repeat it upon calling the program.

Graphics calculators are programmable. In addition, the display shows the data lists (x, y, etc.), for checking and easy modification.

Computer programs include commands . Only the data need be input. In addition, spreadsheet programs (which require Excel) draw graphs automatically and can print out the data, the calculated fits and graphs. This option is the easiest to use and is the most versatile.

All computer programs and spreadsheets are on the CD disks enclosed in this book. Descriptions of the programs and instructions for entering and using the programs are in appendix D. In addition, appendix D describes programs to demonstrate, instruct and perform error analysis.

CALCULATOR: CASIO fX-100S, -115S, -570S, -991S, -85SA, -300SA

PROCEDURE: Entering statistical data.

Set-up for display roundoff. (Permanent: do only on first use.)

Press MODE, MODE , MODE (three times), 1 (FIX), 3, if you wish to display 3 decimals.

Data entry

1. Single variable (mean, standard deviation and standard error of the mean for N entries, one at each value of x).
   a. Press MODE, 2 (SD) to prepare for single variable data entry.
   b. (Optional) To remove all old data, press SHIFT, Mcl or SHIFT, Sci. To check for clearance, press RCL, 3. Display should read 0.000.
   c. Enter data: press x_1, ⌊DATA⌋, x_2, ⌊DATA⌋, ... x_N, ⌊DATA⌋.
2. Single variable (with a frequency or weight f_i at x_i) is the same as part 1, except for
   c. Enter data: x_1, ×, f_1, ⌊DATA⌋, x_2, ×, f_2, ⌊DATA⌋, ... x_N, ×, f_N, ⌊DATA⌋.
3. Two variables (regression calculations)
   a. Press MODE, 3 (LR) to prepare for two variable data entry.
   b. (Optional) To remove all old data, press SHIFT, Mcl or SHIFT, Sci. To check for clearance, press RCL, 3. Display should read 0.000.
   c. For unweighted, single y-value for each x-value, linear regression without error calculation,
   enter data: x_1, ⌊Xo,Yo⌋, y_1, ⌊DATA⌋, x_2, ⌊Xo,Yo⌋, y_2, ⌊DATA⌋, ... x_N, ⌊Xo,Yo⌋, y_N, ⌊DATA⌋.
   d. For linear, exponential, logarithmic, and power law regression with error computation and/or with weights, see the programs below for special forms of input.

CALCULATOR: CASIO fX-100S, -115S, -570S, -991S, -85SA, -300SA

PROCEDURE: Calculation of mean, sample standard deviation (σ_{N-1}), standard error of mean (SEM).

Comments	You enter	Display
First clear statistical registers, then enter single variable data, as shown above in Data Entry section. Sample values are 1, 2, 3.	1. Press 3, ⌊DATA⌋	3.000
	2. Press 2, ⌊DATA⌋	2.000
	3. Press 1, ⌊DATA⌋	1.000
$\sigma_{N-1}=1.0$	4. Press SHIFT, $\bar{x}$	2.000 (mean)
SEM=$(\sigma_{N-1})/\sqrt{(N)}$ = 0.577	5. Press SHIFT, $x\sigma_{N-1}$	1.0 (σ_{N-1})
	6. Press ÷, √, 3, =	.577 (SEM)

CALCULATOR: CASIO FX-100S, -115S, -570S, -991S, -85SA, -300SA.

PROCEDURE: Calculation of mean, sample standard deviation (σ_{N-1}), standard error of mean (SEM) for grouped data, each with weight (frequency) f.

Sample data: x: 1 2 3 4 5

(frequency or weight) f: 4 18 24 18 4 N=68. Mean= 3.000. σ_N =1.000.

SEM=$\sigma_{N-1}/\sqrt{(N)}$= 0.122

Comments	You enter	Display
First clear statistical registers, then enter two variable data for x and y, as shown in Data Entry section (see above).	5, ×, 4, DATA	5.000
	4, ×, 18, DATA	4.000
	. . . . . .	. . . . . .
	1, ×, 4, DATA	1.000
	SHIFT, $\bar{x}$	3.000 (Mean)
	SHIFT, $x\sigma_{n-1}$	1.007 (σ_{n-1})
	÷, √, RCL, 3	68.000 (N=$\sum$ f)
	=	0.122 (SEM)

CALCULATOR: CASIO FX-100S, -115S, -570S, -991S, -85SA, -300SA.

PROCEDURE: Linear regression (unweighted, without errors) uses the built-in calculation.

Comments	You enter	Display
First clear statistical registers, then enter two variable data for x and y, as shown in Data Entry section (see above).	1, ⌊Xo,Yo⌋, 1, ⌊DATA⌋	1.000
	2, ⌊Xo,Yo⌋, 2, ⌊DATA⌋	2.000
	3, ⌊Xo,Yo⌋, 4, ⌊DATA⌋	4.000
Sample data:	4, ⌊Xo,Yo⌋, 7, ⌊DATA⌋	7.000
x: 1 2 3 4	SHIFT, B	2.000 (B)
y: 1 2 4 7	SHIFT, A	-1.500 (A)
y=A+Bx	SHIFT, r	0 .976 (r)
intercept A=-1.500		
slope B=2.000		
correlation coefficient r=0.976		

CALCULATOR: CASIO FX-100S, -115S, -570S, -991S, -85SA, -300SA.

PROCEDURE: Linear regression. Programs to enter weighted data and to find errors in slope B and intercept A. These programs can be used for all four types of regression (linear, log, exponential and power law), which differ slightly in data entry and treatment of errors. Because these programs use weighted data, results will differ from less accurate programs which use unweighted data. Since these calculators lack programmability (i.e., they can't learn a program), you must key in all the steps each time you do a calculation. Each program reduces the problem to a linear regression fit of the form y=A+Bx and builds on the standard, built-in algorithms for A and B. Error algorithms are based on data scatter:

$$\sigma_B = B\sqrt{\frac{\frac{1}{r^2}-1}{N-2}} \quad \text{and} \quad \sigma_A = \sigma_B\sqrt{\frac{\Sigma x^2}{n}}$$

Note that N is the number of x-values, which you will put in by hand and n is the sum of the weights, which the program will insert automatically. You first put in the data with the appropriate weights, if necessary, then run the built in linear regression calculation to get the slope and intercept, run the error computation program (by hand) and then calculate the final errors.

Sample data are the same for all programs:

x: 1 2 3 4

y: 1 2 4 7

To avoid ambiguity in all the programs, A and B refer to intercept and slope on row five of the keyboard; C, D, E, and F refer to storage registers on row 3.

Program Title: Linear regression (continued). Program for straight line, linear regression, $y=A+Bx$, to enter unweighted data and to find errors.

Calculator: CASIO FX-100S, -115S, -570S, -991S, -85SA, -300SA.

Sample values are x, y: 1,1 2,2 3,4 4,7. B=2, A=-1.5, $\sigma_B=0.316$, $\sigma_A=0.866$.

See above for error algorithms.

Press	Display, Sample Value, Comments
MODE, 3 (L.R.), SHIFT, SCI, RCL 3	0.000. Put calculator in linear regression mode, check for clearance of stat registers.
x_1, ⌊Xo,Yo⌋, y_1, ⌊DATA⌋	1.000.
x_2, ⌊Xo,Yo⌋, y_2, ⌊DATA⌋	2.000.
. . . . .	. . . . . .
x_N, ⌊Xo,Yo⌋, y_N, ⌊DATA⌋	7.000.
SHIFT, r, STO, D	0.976. Correlation coefficient r. Save it in D.
RCL, 3, STO, F	4.000. N= No. of data points. Save it in F.
RCL, 1, STO, E	30.000. $\sum x^2$. Save it in E.
SHIFT, A (button on row 5)	-1.500. Intercept A. Write it down.
SHIFT, B (button on row 5), STO, C	2.000. Slope B. Write it down and save it in C.
MODE, 1 (COMP)	Computation mode.
RCL, C , ×, √, (, (, RCL, D, 1/x, x^2, -, 1,)	0.050. Intermediate result.
÷, (, N, - , 2,),), =	0.316. σ_B. N is no. of data points (4 here).
×, √, (, RCL, E, ÷, RCL, F,)	7.5= $\sum x^2/N$.
=	0.866. σ_A

CALCULATOR: CASIO FX-100S, -115S, -570S, -991S, -85SA, -300SA

PROCEDURE: Logarithmic regression. This fits data to a logarithmic curve $y=A+B\ln(x)$, such as the potential between two cylindrical conductors. Sample values are x, y: 1,1 2,2 3,4 4,7. B=4.023, A=0.304, σ_B=1.262, σ_A=1.199. See above for error algorithms. Since there is no weight factor, all calculators give the same fit.

Press	Display, Sample Value, Comments
MODE, 3 (L.R.), SHIFT, SCI, RCL 3	0.000. Put calculator in linear regression mode, check for clearance of stat registers.
ln, x_1, =, ⌊Xo,Yo⌋, y_1, ⌊DATA⌋	1.000.
ln, x_2, =, ⌊Xo,Yo⌋, y_2, ⌊DATA⌋	2.000.
.....	..... .
ln, x_N, =, ⌊Xo,Yo⌋, y_N, ⌊DATA⌋	7.000.
SHIFT, r, STO, D	0.914. Correlation coefficient r. Save it in D.
RCL, 3, STO, F	4.000. N= No. of data points. Save it in F.
RCL, 1, STO, E	3.609. $\sum x^2$. Save it in E.
SHIFT, A (button on row 5)	0.304. Intercept A. Write it down.
SHIFT, B (button on row 5), STO, C	4.023. Slope B. Write it down and save it in C.
MODE, 1 (COMP)	Computation mode.
RCL, C , ×, √, (, (, RCL, D, 1/x, x^2, -, 1,)	0.197. Intermediate result.
÷, (, N, - , 2,),), =	1.262. σ_B. N is no. of data points (4 here).
×, √, (, RCL, E, ÷, RCL, F,)	0.902= $\sum x^2/N$.
=	1.199. σ_A.

CALCULATOR: CASIO FX-100S, -115S, -570S, -991S, -85SA, -300SA

PROCEDURE: Exponential regression (weighted, with errors). This fits data to an exponentially varying curve $y=ae^{Bx}=e^{A+Bx}$, such as the voltage in an RC or LR circuit. It uses weights=y^2 to transform the problem into linear regression with the equation ln(y)=ln(a)+Bx, with intercept A=ln(a) and slope B. Sample values are x, y, w: 1,1,1; 2, 2, 4; 3, 4, 16; 4, 7, 49; B=0.611, A= -0.488, σ_B=0.030, σ_A=0.109, a=e^A=0.614, σ_a= a σ_A=0.067. Unweighted fits will deviate from these results.

Press	Display, Sample Value, Comments
MODE, 3 (L.R.), SHIFT, SCI, RCL 3	0.000. Put calculator in linear regression mode, check for clearance of stat registers.
x_1, ⌊Xo,Yo⌋, ln, y_1, ×, y_1^2, ⌊DATA⌋	0.000.
x_2, ⌊Xo,Yo⌋, ln, y_2, ×, y_2^2, ⌊DATA⌋	0.693.
.....	....
x_N, ⌊Xo,Yo⌋, ln, y_N, ×, y_N^2, ⌊DATA⌋	1.946.
SHIFT, r, STO, D	0.998. Correlation coefficient . Save it in D.
RCL, 3, STO, F	70.000. n=sum of weights (n≠N!). Save it in F.
RCL, 1, STO, E	945.000. $\sum wx^2$. Save it in E.
SHIFT, A (button on row 5), SHIFT, e^x, =	-0.488. Intercept A. 0.614. a=e^A. Record it.
SHIFT, B (button on row 5), STO, C	0.611. Slope B. Write it down and save in C.
MODE, 1 (COMP)	Computation mode.
RCL, C , ×, √, (, (, RCL, D, 1/x, x^2, -, 1,)	0.005. Intermediate result.
÷, (, N, - , 2,),), =	0.030. σ_B. N is no. of data points (N=4 ≠n!).
×, √, (, RCL, E, ÷, RCL, F,)	13.5= $\sum x^2/n$.
=	0.109. σ_A
×, a, =	0.067. σ_a= a σ_A
Time constant τ=1/B=1.64. Error=$\tau\sigma_B$/B=0.08	

CALCULATOR: CASIO FX-100S, -115S, -570S, -991S, -85SA, -300SA

PROCEDURE: Exponential regression (counts,weighted, with errors). This fits data to an exponentially varying curve $y=ae^{Bx}=e^{A+Bx}$, such as counting experiments for radioactive decay or bacterial growth. It uses weights=y to transform the problem into linear regression with the equation ln(y)=ln(a)+Bx, with intercept A=ln(a) and slope B. Sample values are x, y, w: 1,1,1; 2, 2, 2; 3, 4, 4; 4, 7, 7; B=0.634, A= -0.569, σ_B=0.028, σ_A=0.092, a=e^A=0.566, σ_a= aσ_A= 0.052. Unweighted fits will deviate from these results.

Press	Display, Sample Value, Comments
MODE, 3 (L.R.), SHIFT, SCI, RCL 3	0.000. Put calculator in linear regression mode, check for clearance of stat registers.
x_1, ⌊Xo,Yo⌋, ln, y_1, ×, y_1, ⌊DATA⌋	0.000.
x_2, ⌊Xo,Yo⌋, ln, y_2, ×, y_2, ⌊DATA⌋	0.693
.....	....
x_N, ⌊Xo,Yo⌋, ln, y_N, ×, y_2, ⌊DATA⌋	1.946.
SHIFT, r, STO, D	0.998. Correlation coefficient . Save it in D.
RCL, 3, STO, F	14.000. n= sum of weights (n≠N!). Save in F.
RCL, 1, STO, E	157.000. $\sum x^2$ (weighted sum). Save it in E.
SHIFT, A (button on row 5), SHIFT, e^x, =	-0.569. Intercept A. 0.566. a=e^A. Record it.
SHIFT, B (button on row 5), STO, C	0.634. Slope B. Write it down and save in C.
MODE, 1 (COMP)	Computation mode.
RCL, C , ×, √, (, (, RCL, D, 1/x, x^2, -, 1,)	0.004. Intermediate result.
÷, (, N, - , 2,),), =	0.028. σ_B. N is no. of data points (N=4 ≠n!).
×, √, (, RCL, E, ÷, RCL, F,)	11.214= $\sum x^2/n$.
=	0.092. σ_A
×, a, =	0.052. σ_a= aσ_A

Dividing time $T_{1/2}$=ln(2)/B=1.09 Error= $T_{1/2}\sigma_B$/B=0.05

CALCULATOR: CASIO FX-100S, -115S, -570S, -991S, -85SA, -300SA

PROCEDURE: Power law regression. This fits data to a power law curve $y=ax^B$ such as the area or volume of a sphere as a function of its radius. It uses weights=y^2 to transform the problem into linear regression with the equation ln(y)=ln(a)+Bx, with intercept A=ln(a) and slope B. Sample values are x, y, w: 1,1,1; 2, 2, 4; 3, 4, 16; 4, 7, 49; B=1.659, A= -0.374, σ_B=0.175, σ_A=0.235, a=e^A=0.688, $\sigma_a= a\sigma_A$=0.155. Unweighted fits will deviate from these results.

Press	Display, Sample Value, Comments
MODE, 3 (L.R.), SHIFT, SCI, RCL 3	0.000. Put calculator in linear regression mode, check for clearance of stat registers.
ln, x_1, =, $\lfloor$Xo,Yo$\rfloor$, ln, y_1, ×, y_1^2, $\lfloor$DATA$\rfloor$	0.000.
ln, x_2, =, $\lfloor$Xo,Yo$\rfloor$, ln, y_2, ×, y_2^2, $\lfloor$DATA$\rfloor$	0.693.
. . . . .	. . . .
ln, x_N, =, $\lfloor$Xo,Yo$\rfloor$, ln, y_N, ×, y_N^2, $\lfloor$DATA$\rfloor$	1.946.
SHIFT, r, STO, D	0.989. Correlation coefficient . Save it in D.
RCL, 3, STO, F	70.000. n= sum of weights (n≠N!). Save in F.
RCL, 1, STO, E	115.402. ΣX^2 (weighted sum). Save in E.
SHIFT, A (button on row 5), SHIFT, e^x, =	-0.374. Intercept A. 0.688. a=e^A. Record it.
SHIFT, B (button on row 5), STO, C	1.659. Slope B. Write it down and save in C.
MODE, 1 (COMP)	Computation mode.
RCL, C , ×, √, (, (, RCL, D, 1/x, x^2, -, 1,)	0.022. Intermediate result.
÷, (, N, - , 2,),), =	0.175. σ_B. N is no. of data points (N=4 ≠n!).
×, √, (, RCL, E, ÷, RCL, F,)	1.649= $\Sigma x^2/n$.
=	0.225. σ_A
×, a, =	0.155. $\sigma_a = a\,\sigma_A$

Trigonometric Operations on Casio Calculators.

This calculator's mode (deg, rad, grd) is shown on the display. To change it, press MODE twice, then enter 1, 2, or 3 to get the desired mode.

1. How to make conversions from one set of units to another:

1a. Decimal degrees D.d to degrees, minutes and seconds D° M' S".

SHIFT, [° ' "] . (See Example below, line 2.)

1b. Degrees, minutes, and seconds D° M' S" to decimal degrees D.d .

[° ' "] (See Example, line 3.) To enter an angle given in D° M' S", press D, +, M, ÷, 60, +, S, ÷, 3600, =. Angle in D.d will be in display.

1c. Decimal degrees D.d to radians R.

SIN, D.d, =, MODE, MODE, 2 (RAD), SHIFT, SIN^{-1}, ANS, =. (See Example below, lines 4, 5.)

1d. Radians R to decimal degrees D.d .

SIN, R, =, MODE, MODE, 1 (DEG), SHIFT, SIN^{-1}, ANS, =. (See Example below, line 6.)

2. How to find the sine, cosine, or tangent of an angle θ,

2a. given in degrees, minutes, and seconds D° M' S" or in decimal degrees.

MODE, MODE, 1(DEG), θ (enter as D.d as shown in 1b), SIN, =.

2b. Given in radians R.

MODE, MODE, 2 (RAD), SIN, θ (in radians), =.

3. Given the sine, cosine, or tangent of an angle θ, to find θ (to find $\sin^{-1}(x)$, $\cos^{-1}(x)$, or $\tan^{-1}(x)$).

3a. in radians R.

MODE, MODE, 2 (RAD), SHIFT, SIN^{-1} (or COS^{-1} or TAN^{-1}), x, =. (See Example, line 5.)

3b. in decimal degrees D.d .

MODE, MODE, 1 (DEG), SHIFT, SIN^{-1} (or COS^{-1} or TAN^{-1}), x, =. (See Example, line 4.)

3c. in degrees, minutes, and seconds D ° M' S". Follow 3b, then 1a.

Example for Casios (with calculator fixed to 8 decimals).

Convert 1 7/8° to degrees, minutes, and seconds and back again. Take the sine and then get back to the original angle. Convert the angle to radians and back to decimal degrees.

Solution.

Step	Press	Reading	Comments
1	1, +, 7, ÷, 8, =	1.87500000	1.875°, the angle in dec. deg.
2	SHIFT, [° ' "]	1°52°30.	1°52'30", the angle in D° M' S"
3	[° ' "]	1.87500000	1.875°, the angle in dec. deg.
4	SIN, =	0.03271908	sin(1.875°)
5	MODE, MODE, 2 (RAD), SHIFT, SIN^{-1}, =	0.03272492	The angle in radians
6	SIN, =, MODE, MODE, 1(DEG), SHIFT, SIN^{-1}, =	1.87500000	1.875°, the angle in dec. deg.

*NOTE: This calculator has no direct way of converting degrees to radians and *vice versa*. The method used here is to calculate the sine of the angle, switch modes and calculate the inverse sine. The display shows "RAD" or "DEG", depending on whether it is in radian or degree mode. See the Example for using these keys.

CALCULATOR: CASIO fx-7400G (9850G with minor changes)

PROCEDURE: Entering statistical data.

Note: There are three different right arrows: ► (on row one), ▷(on REPLAY) and -> (on row 5) . Make sure you understand the difference among these symbols.

Set-up for display roundoff and list names. (Permanent: do only on first use.)

In any mode from 1-3, press **SHIFT** **SET UP.** Move to **Display** , then choose the number of decimals you wish to display. For example, press **F1**, then **F4** for 3 decimals or ►, **F1** for 4 decimals, etc. Display will read Fix 3 for 3 decimals, Fix 4 for 4 decimals, etc. Press **QUIT**.

Set-up for data lists. Press **MENU**, 2 (or move to STAT and press **EXE**) Data lists will

show. Check the default set-up by pressing **F2**(CALC), **F4**(SET). The display should read

1Var X :List 1

1Var F :1

2Var X :List1

2Var Y :List2

2Var F :1,

which says that the X variables are in List 1 with weight=1 (each data point counted once) for single variable statistics. For two variable statistics, such as regression, X is also in List 1, Y is in list 2, and the weights are 1 (each point counted once.) If the display is wrong, move to the line that needs correction and press the F-key to give the correct reading. Normally, this set-up will serve your needs. If not, the program here will tell you which set-up to use.

3. Press **QUIT** to return to the lists and enter data. You are now ready for calculations.

4. Begin data entry (see next line).

Data entry

1. (optional) To remove old data, go to the cell where you wish to remove the datum and press

   ► (on row one), F1 (DEL). To remove a whole column, go to the top (LIST 1, LIST 2, etc.) and press F2 (DEL-A). You can skip this step, go to step 2 and overwrite old data.

2. Enter data x_1, **EXE**, x_2, **EXE**, etc., until all x entries are in first column. Repeat in second

   column for y-entries for two-variable data. If necessary, enter further data in third or fourth column for frequency if needed. Proceed to statistical calculations (see below).

CALCULATOR: CASIO FX-7400G (9850G WITH MINOR CHANGES)

PROCEDURE: Calculation of mean, sample standard deviation (σ_{N-1}), standard error of mean (SEM),

Comments	You enter	Display
Enter single variable data in List 1.	1. Press **F2**, (CALC), **F1** (1VAR).	1-Variable
		$\bar{x}$ = 2
Sample values are 1, 2, 3.		Σx= 6
		Σx^2= 14
	2. Press ▽, ▽	xσn=0.81649
		xσn-1= 1
	3. **MENU**, 1 (RUN) , 1 **÷**	n= 3
SEM=(σ_{N-1})/√(N) = 0.5774	**SHIFT**, √, 3, **EXE**	1÷ √3
		0.5774

CALCULATOR: CASIO FX-7400G (9850G WITH MINOR CHANGES)

PROCEDURE: Calculation of mean, sample standard deviation (σ_{N-1}), standard error of mean (SEM) for grouped data.

Comments	You enter	Display
Enter single variable data,	1. Press **QUIT**, **F2** (CALC),	1-Variable
Sample values are shown as entered for x and frequency F.	**F1** (1VAR)	$\bar{x}$ = 3
List 1(x) List 2(F)		Σx= 204
1 4		Σx^2= 680
2 18		xσn= 1
3 24	2. Press ▽, ▽	xσn-1=1.00743
4 18		n= 68
5 4		
Go to set-up for data lists (see above) and change second line to		
1Var F :List 2.		
Then quit or exit edit menu.	3. **MENU**, 1 (RUN) , 1.007	
SEM=(σ_{N-1})/√(N) = 0.1221	**÷**, **SHIFT**, √, 68, **EXE**	1.007÷ √68
		.1221

CALCULATOR: CASIO FX-7400G (9850G WITH MINOR CHANGES)

PROCEDURE: Linear regression (unweighted, without errors)

Comments	You enter	Display
Enter with this sample data: List 1(x) List 2(y) 1 1 2 2 3 4 4 7 Note: Use of a and b is reversed from usual meaning. Be careful! Go to setup menu and restore it to default value (see "Entering Statistical Data" above) if necessary.	MENU, 2 (STAT, or go to STAT and press EXE, F2 (CALC), F3 REG, F1 (X)	LinearReg a= 2 b= -1.5 r= 0.9759 y=ax+b

For other forms of regression without weighting or errors, follow the same procedure as above, except for the last step. Instead of F1(X), press ►, then as follows:

Logarithmic regression: press F1 (Log). y=a+b ln(x). a=0.30378; b=4.02286

Exponential regression: press F2 (Exp). $y=ae^{bx}$. a=0.53452; b=0.65308

Power Law regression: press F3 (Pwr). $y=ax^{b}$. a=0.9094; b=1.38613

Display will show results for each calculation as shown.

CALCULATOR: CASIO FX-7400G (9850G with minor changes)

PROCEDURE: Linear regression. Program to find errors in a and b. This program, with small additions, can be used for all four types of regression (linear, log, exponential, and power law).

Comments	You enter	Display
Before you can find the errors you must type in the data, these programs, and do the regression calculation. Instructions for use of these programs are shown later for each type of calculation. Name the prgms "LRERRORS" and "ERRORINT". Enter the programs. Use QUIT and/or MENU to get out of editing.	1. MENU, 6 (**PRGM,** or move to PGRM and press EXE), F3 (NEW),L, R, E, R, R, O, R, S, EXE, ►, F2 ("), ALPHA, N, F2 ("), SHIFT, PRGM, ►, F1(?), ->, ALPHA, N, ►, F2 (:), VARS, ►, F1 (STAT), F3 (GRPH), F1 (a), SHIFT, √, (, (, F4 (r), x^2, SHIFT, x^{-1}, -, 1,), ÷, (ALPHA, N, -, 2,),), ->, ALPHA, E, QUIT	= LRERRORS = "N"?->N:a√((r^2 $^{-1}$-1)÷ (N-2))->E
	2. MENU 6 (**PRGM,** or move to PGRM and press EXE), F3 (NEW), E, R, R, O, R, I, N, T, EXE, ALPHA, E, SHIFT, √, (, VARS, ►, F1 (STAT), F1 (x), F4 (Σx^2, ÷, F1 (n),), QUIT	= ERRORINT= E√(Σx^2÷n)

CALCULATOR: CASIO FX-7400G (9850G WITH MINOR CHANGES)

PROCEDURE: Regression calculations with weighted data and errors.

General procedure:

1. Enter x-data in column 1 and y-data in column 2. Sample data are shown below.
2. Choose type of regression calculation (linear, log, exponential, power law) and follow instructions given below for it.
3. Run error programs "LRERRORS" and "ERRORINT" (which you had entered) to find errors.
4. Use these values to calculate final errors. For the sample data and linear regression y=b+ax, the values are for the difference quotient a=2, σ_a=0.316; for the intercept b=-1.5; σ_b=0.866.

Linear Regression (unweighted, with errors)

Comments	You enter	Display
Enter with this sample data: List 1(x) List 2(y) 1 1 2 2 3 4 4 7	[MENU], 2 (or move to STAT and press [EXE]), F2 (CALC), F3 (REG), F1 (x)	LinearReg a= 2 b= -1.5 r= 0.9759 y=ax+b
Setup for data lists should be the default. (See "Entering Statistical Data" above.)	[MENU], 6 (or move to PRGM and press [EXE]), move to LRERRORS, F1 (EXE)	N?
Run prog. N=no. of data points. Display shows the error in a, which is 0.3162.	N, [EXE]	4 .3162
	[MENU], 2 (or move to STAT and press [EXE]), F2 (CALC), F1 (1VAR)	1-Variable $\bar{x}$ = 2.5 Σx= 10, etc.
Error for the intercept b is 0.8660.	[MENU], 6 (or move to PRGM and press [EXE]), move to ERRORINT, [EXE])	.8660

CALCULATOR: CASIO FX-7400G (9850G WITH MINOR CHANGES)

PROCEDURE: Logarithmic regression. This fits data to a logarithmic curve, such as the potential between two cylindrical conductors. The function is of the form y=b+a ln(x). For the sample data b=0.30378; σ_b=1.1988; a=4.02286; σ_a=1.2621. There is no weight factor. Except for the first few keystrokes, which change List 1 from x to ln(x), the set-up programs and keystrokes are the same as linear regression, the preceding program.

Comments	You enter	Display
Enter with this sample data:	[MENU], 1, [OPTN], [ln], F1	ln List 1->Lis
List 1(x) List 2(y)	(LIST), F1, (LIST),1, ->, F1,	t 1
1 1	(LIST), 1, [EXE]	Done
2 2	[MENU], 2 (or move to STAT	List 1(x) List 2(y)
3 4	and press [EXE])	0 1
4 7		0.6931 2
Setup for data lists should be the default. (See "Entering Statistical Data" above.)		1.0986 4
	F2 (CALC),	1.3862 7
Run prog. N=no. of data points. Display shows the error in a.	F3 (REG), F1 (x)	Linear Reg
		a= 4.02286
		b= 0.30378
		....
	[MENU], 6 (or move to PRGM	
	and press [EXE]), move to	
	LRERRORS, F1 (EXE)	N?
Error in a=1.2621	N, [EXE]	4
		1.2621
	[MENU], 2 (or move to STAT	
	and press [EXE]), F2 (CALC),	
	F1 (1VAR)	1-Variable
		$\overline{x}$ = 0.79451
		....
	[MENU], 6 (or move to PRGM	
	and press [EXE]), move to	
Error in b=1.1988	ERRORINT, [EXE])	
		1.1988

CALCULATOR: CASIO FX-7400G (9850G with minor changes)

PROCEDURE: Exponential regression. This fits data to an exponentially varying curve, such as the voltage in an RC- or LR-circuit. The function is of the form $y=be^{ax}$. A weight factor proportional to the square of the voltage in List 3 corrects for the large relative errors for small signals.

Comments	You enter	Display
Enter with this sample data: List 1(x) List 2(y) 1 1 2 2 3 4 4 7	**MENU**, 2 (or move to STAT and press **EXE**, F2 (CALC), F4 (SET), and set as follows; 1 Var X :List1, 1 Var F:List 3, 2VarX: List 1, 2VarY:List 2, 2Var F:List 3. Press **QUIT**	
	MENU, 1, **OPTN**, F1 (LIST), F1 (LIST), 2, x^2, ->, F1 (LIST), 3, **EXE**	List 2^2->List 3 Done
For final answers see below.	ln, F1 (LIST), 2, **->**, F1 (LIST), 2, **EXE**	ln List 2->Lis t2 Done
	MENU, 2 (or move to STAT and press EXE)	List 1 List 2 List 3 1 0 1 2 0.6931 4 3 1.3862 16
difference quotient a=0.61063 intercept b=0.48839	F2 (CALC), F3 (REG), F1 (x)	LinearReg a= 0.61063 b=-0.48839
σ_a = 0.0297	MENU, 6 (or move to PRGM and press EXE), move to LRERRORS, F1 (EXE) N, EXE	N? 4 0.0297
	MENU, 2 (or move to STAT and press EXE), F2 (CALC), F1 (1VAR)	1-Variable $\bar{x}$ =3.61428···
σ_b = 0.1090	MENU, 6 (or move to PRGM and press EXE), move to ERRORINT, EXE)	0.1090

Final answers for exponential regression for $y=Be^{Ax}$ from b, a, σ_b, σ_a, with examples:

Values: A=a=0.61063 $B=e^b$=0.6136

Errors: $\sigma_A=\sigma_a$= 0.0297 $\sigma_B=B\sigma_b$=0.067

Time constant=τ=1/A=1.64. Error= $\tau\,\sigma_A$/A=0.08

CALCULATOR: CASIO FX-7400G (9850G WITH MINOR CHANGES)

PROCEDURE: Exponential regression (counts). This fits data to an exponentially varying count rate, such as the decay of a radioactive isotope or the growth of a cell culture. A weight factor proportional to the count rate corrects for the large relative errors for small signals.

Comments	You enter	Display
Enter with this sample data: List 1(x) List 2(y) 1 1 2 2 3 4 4 7	**MENU**, 2 (or move to STAT and press **EXE**, F2 (CALC), F4 (SET), and set as follows; 1 Var X :List1, 1 Var F:List 3, 2VarX: List 1, 2VarY:List 2, 2Var F:List 3. Press **QUIT**	
	MENU, 1, **OPTN**, F1 (LIST), F1 (LIST), 2, **->**, F1 (LIST), 3, **EXE**	List 2->List 3 Done
For final answers see below.	ln, F1 (LIST), 2, **->**, F1 (LIST), 2, **EXE**	ln List 2->Lis t 2 Done
	MENU, 2 (or move to STAT and press **EXE**)	List 1 List 2 List 3 1 0 1 2 0.6931 2 3 1.3862 4
Difference quotient a=0.63371 Intercept b=0.56887	F2 (CALC), F3 (REG), F1 (x)	LinearReg a= 0.63371 b=-0.56887
	MENU, 6 (or move to PRGM and press **EXE**), move to LRERRORS, F1 (EXE)	N? 4
$\sigma_a = 0.0276$	N, **EXE**	0.0276
	MENU, 2 (or move to STAT and press **EXE**), F2 (CALC), F1 (1VAR)	1-Variable $\bar{x}$ =3.21428···
$\sigma_b = 0.0923$	**MENU**, 6 (or move to PRGM and press **EXE**), move to ERRORINT, **EXE**)	0.0923

Final answers for exponential regression for $y=Ae^{Bx}$ from a, b, ERRA, and ERRB, with examples:

Values: A=a=0.634 $\quad$ $B=e^{b}=0.566$

Errors: $\sigma_A=\sigma_a= 0.028$ $\quad$ $\sigma_B=B\ \sigma_b=0.052$

Cell division time=$T_{1/2}$= ln(2)/A=1.09. Error=$T_{1/2}\ \sigma_A/A$=0.05. The same expressions hold for radioactive half-life, except that (-B) is used instead of B.

CALCULATOR: CASIO FX-7400G (9850G with minor changes)

PROCEDURE: Power law regression. This fits data to an power law relation, such as the mass of a ball as a function of its radius. A weight factor proportional to the square of y corrects for the large relative errors for small values.

Comments	You enter	Display
Enter with this sample data: List 1(x) List 2(y) 1 1 2 2 3 4 4 7	MENU, 2 (or move to STAT and press EXE, F2 (CALC), F4 (**SET**), and set as follows; 1 Var X :List1, 1 Var F:List 3, 2VarX: List 1, 2VarY:List 2, 2Var F:List 3	
	MENU, 1, OPTN, F1 (LIST), F1 (LIST), 2, x^2, ->, F1 (LIST), 3, EXE	List 2^2->List 3 Done
For final answers see below.	ln, F1 (LIST), 1, ->, F1 (LIST), 1, EXE	ln List 1->Lis t 1 Done
	ln, F1 (LIST), 2, ->, F1 (LIST), 2, EXE	ln List 2->Lis t 2 Done
	MENU, 2 (or move to STAT *and press* EXE)	List 1 List 2 List 3 0 0 1 0.6931 0.6931 4 1.0986 1.3862 16
	F2 (CALC), F3 (REG), F1 (x)	
difference quotient a=1.65933 intercept b=-0.37402		LinearReg a= 1.65933 b=-0.37402
	MENU, 6 (or move to PRGM and press EXE), move to LRERRORS, F1 (**EXE**)	
	N, EXE	N? 4
$\sigma_a = 0.1749$	MENU, 2 (or move to STAT and press EXE), F2 (CALC), F1 (1VAR)	0.1749
	MENU, 6 (or move to PRGM and press EXE), move to ERRORINT, EXE	1-Variable $\overline{x}$ =1.26112···
$\sigma_b = 0.2246$		0.2246

Final answers for power law regression for $y=Bx^A$ from a, b, σ_A, σ_B, with examples.

Values: A=a= 1.659 $B=e^b=0.688$

Errors: $\sigma_A=\sigma_a= 0.175$ $\sigma_B=B\sigma_b=0.155$

Trigonometric Operations on the Casio fx-7400G Calculator.

To view or change calculator's mode (deg, rad, grd), press MENU, 1 (Run), **SHIFT**, **SET UP**, move to Angle and, if a change is desired, press F1 (Deg) or F2 (Rad). Then **QUIT**.

1. How to make conversions from one set of units to another:

1a. Decimal degrees D.d to degrees, minutes, and seconds D° M' S".

In RUN and degree mode, enter D.d, Press **OPTN**, ►, F2 (ANGL), ►, F1 (° ' "), **EXE**, F2 (°' "). (See Example below, line 2.)

1b. Degrees, minutes and seconds D° M' S" to decimal degrees D.d.

To enter an angle given in D° M' S", in run mode press **OPTN**, ►, F2 (ANGL), ►, D°, F1(° ' "), M, F1 (° ' "), S, F1 (° ' "), **EXE**

(See Example below, line 3.)

1c. Decimal degrees D.d to radians R.

Put calculator in RAD and RUN mode (see above), enter D.d, **OPTN**, ►, F2 (ANGL), F1(°), **EXE** (See Example, line 6.)

1d. Radians R to decimal degrees D.d.

Put calculator in DEG and RUN mode (see above), enter R, **OPTN**, ►, F2 (ANGL), F2(r), **EXE**. (See Example, line 7.)

2. How to find the sine, cosine or tangent of an angle θ:

2a. given in degrees, minutes, and seconds D° M' S" or in decimal degrees.

In Deg and RUN mode, press **SIN**, θ (enter as D.d or D.M.S as shown in 1b), **EXE** (See Example, line 4.)

2b. Given in radians R.

In Rad and RUN mode, press **SIN**, θ (in radians), **EXE**

3. Given the sine, cosine, or tangent x of an angle θ, to find θ (to find $\sin^{-1}(x)$, $\cos^{-1}(x)$, or $\tan^{-1}(x)$).

3a. in radians R.

In Rad and RUN mode, **SHIFT**, SIN^{-1} (or COS^{-1} or TAN^{-1}), x, **EXE**.

3b. in decimal degrees D.d.

In Deg and RUN mode, **SHIFT**, SIN^{-1} (or COS^{-1} or TAN^{-1}), x, EXE. (See Example, line 5.)

3c. in degrees, minutes, and seconds D ° M' S". Follow 3b, then **OPTN**, ►, F2 (ANGL), ►, F1 (° ' "), **EXE**, F2 (°' ").

Example for Casio fx-7400G (with calculator fixed to 8 decimals, in RUN and degree mode).

Convert 1 7/8° to degrees, minutes and seconds and back again. Take the sine and then get back to the original angle. Convert the angle to radians and back to decimal degrees.

Solution.

Step	Press	Reading	Comments
1	1, +, 7, ÷, 8, **EXE**	1.87500000	1.875°, the angle in dec. deg.
2	**OPTN**, ►, F2 (ANGL), ►, F1 (° ' "), **EXE**, F2 (°' ").	1°52'30"	the angle in D° M' S"

3	EXE	1.87500000	1.875°, the angle in dec. deg.
4	SIN, SHIFT, Ans, EXE	0.03271908	sin(1.875°)
5	SHIFT, $\sin^{-1}$, SHIFT,		
	Ans, EXE	1.87500000	1.875°, the angle in dec. deg.
6	SHIFT SET UP, move to		
	Angle, F2 (Rad), QUIT,		
	OPTN, ►, F2 (ANGL),		
	F1(°), EXE	0.03272492	The angle in radians
7	**SHIFT** SET UP, move to		
	Angle, F1 (Deg), QUIT,		
	OPTN, ►, F2 (ANGL),		
	F2(r), EXE	1.87500000	1.875°, the angle in dec. deg.

Common Problems with Casio graphing Calculators

Problem	Solution
Displays wrong key code (alphabetical *vs.* non-alphabetical)	Press ALPHA (toggles alphabetical codes on or off).
F1, F2, F3, or F4 button doesn't correspond to the right word on bottom line of display.	Press ► (on first row) to get a new bottom line of display.
Calculator won't enter my command or won't take me where I want to go.	Press EXE or REPLAY: △, ▽, ▷, or ◁.
I can't escape from where I am.	Press QUIT, AC/ON, or MENU.

CALCULATOR: HP-20S

PROCEDURE: Entering statistical data.

Setup for display roundoff. (Permanent: do only on first use.)

Press [↰], FIX 3 if you wish to display 3 decimals.

Data entry

1. (optional) To remove old data, press [↱]CLΣ.
2. Single variable [mean, standard deviation, and standard error of the mean (SEM) for N entries, one at each value of x.] Enter data x_1, Σ+, x_2, Σ+, ··· x_N, Σ+.
3. Two-variables. a. For mean, standard deviation and standard error of the mean for N values of x, with y entries at each value of x. Also for linear regression without error calculation. Enter data x_1, INPUT, y_1, Σ+, x_2, INPUT, y_2, Σ+, ··· x_N, INPUT, y_N, Σ+.

   b. For linear, exponential, logarithmic and power law regression with error computation, see the programs for special forms of input.

CALCULATOR: HP 20S

PROCEDURE: Calculation of mean, sample standard deviation (σ_{N-1}), standard error of mean (SEM).

Comments	You enter	Display
First clear statistical registers, then enter single variable data, as shown in Data Entry section (see above).	1. Press 3, Σ+	1.000 (N=1)
	2. Press 2, Σ+	2.000 (N=2)
Sample values are 1, 2, 3.	3. Press 1, Σ+	3.000 (N=3)
	4. Press [↱], $\bar{x}$, $\bar{y}$	2.000 (Mean)
σ_{N-1}= 1.0	5. Press [↱], σ_x,σ_y	1.0 (σ_{N-1})
SEM=(σ_{N-1})/√(N) = 0.577	6. Press ÷, 3, √x, =	.577 (SEM)

CALCULATOR: HP-20S.

PROCEDURE: Calculation of mean, sample standard deviation (σ_{N-1}), standard error of mean (SEM) for grouped data, each with weight (frequency) y.

Sample data: x: 1 2 3 4 5

(frequency or weight) y: 4 18 24 18 4 N=68. Mean= 3.000. σ_N =1.000.
$SEM=\sigma_N/\sqrt{(N-1)}$= 0.122

Calculation of σ_N: $\sigma_N^2 = (\Sigma yx^2)/N - (\Sigma yx/N)^2$. The second term is $\bar{x}^2$, obtainable from data with x and y as the variables. The first term must be obtained from a second set of data with x^2 and y as the variables.

Comments	You enter	Display
First clear statistical registers, then enter two variable data for x and y, as shown in Data Entry section (see above). Then repeat the process for x^2 and y.	5, INPUT, 4, Σ+	1.000 (first data)
	4, INPUT, 18, Σ+	2.000 (second data)
	. . . . . .	. . . . . .
	1, INPUT, 4, Σ+	5.000 (fifth data)
	[↱], $\bar{x}_w$	3.000 (Mean= (Σyx/N)
	[↱], CLΣ	Clear statistical registers
	25, INPUT, 4, Σ+	1.000 (first data for x^2, y)
	16, INPUT, 18, Σ+	2.000 (second data for x^2, y)
	. . . . .	. . . . .
	4, INPUT, 18, Σ+	4.000 (fourth data for x^2,y)
	1, INPUT, 4, Σ+	5.000 (fifth data for x^2, y)
	[↱], $\bar{x}_w$	10.000 $(\Sigma yx^2)/N$
	(, 10, -, 3, [↱], x^2,), √	1.000 (σ_N)
	÷, 67, √x, =	0.122 (SEM)

CALCULATOR: HP20S

PROCEDURE: Linear regression (unweighted, without errors)

Comments	You enter	Display
First clear statistical registers, then enter two variable data for x and y, as shown in Data Entry section (see above).	1, INPUT, 1, Σ+	1.000 (first data)
	2, INPUT, 2, Σ+	2.000 (second data)
	3, INPUT, 4, Σ+	3.000 (third data)
Sample data:	4, INPUT, 7, Σ+	4.000 (fourth data)
x: 1 2 3 4	[↱], m,b	2.000 (m)
y: 1 2 4 7	[↰], SWAP	-1.500 (b)
y=mx+b		
intercept b=-1.500		
slope m=2.000		
correlation coefficient r=0.976		

CALCULATOR: HP-20S.

PROCEDURE: Linear regression. Program to enter unweighted data and to find errors in slope m and intercept b. This program, with small additions, can be used for all four types of regression (linear, log, exponential and power law). Because this program does not weight data, its results will differ from programs that used weighted data.

Comments	You enter	Display
Before you can find the errors you must load and trim "fit", enter this program, the data, and do the regression calculation.	[↰], PRGM, [↰], LOAD, F	Fit (momentarily), then 00- (loads "fit" program)
First load the program "fit".	[↰], GTO, . , 33	33- 51 42 2
To make space for the error computation, remove steps 21-33 of "fit".	←←←←←←←←←←←←←(13 times)	20- 51 41 1
	[↰], GTO, . , 64	64- 51 31 (last line of "fit" program)
Key in the error computation subprogram in steps 65-93. The added program and instructions for each type of regression calculation are shown below.	Error computation subprogram is shown below.	

Program Title: Linear Regression (continued). Program to enter unweighted data and to find errors in m and b.

Calculator: HP20S. Sample values are from straight line, linear regression , y=mx+b.

x, y: 1,1 2,2 3,4 4,7. m=2, b=-1.5, σ_m=0.316, σ_b=0.866. Errors are based on scatter of data:

$$\text{(slope) } \sigma_m = \sqrt{\frac{\left(\frac{1}{r^2} - 1\right)}{N-2}} = \frac{\tan(\text{acos}(r))}{\sqrt{N-2}}, \text{ (intercept) } \sigma_b = \sigma_m \sqrt{\frac{\Sigma x^2}{N}}.$$

Location	Code	Key	Comments, Sample Value
00	00-		
1-64			Remainder of "Fit" program after editing.
65	61 26	⮫, RTN	Ends "fit" program.
66	61 41 6	⮫, LBL,6	Label 6. Beginning of error computation subprogram.
67	61 16	⮫, m,b	Get the slope m. Sample value 2.000
68	21 3	STO, 3	Save slope m and store it in location 3.
69	71	C	Clear the display and pending operations.
70	61 15	⮫, ŷ,r	Gets correlation coefficient in next step.
71	51 31	⮪, SWAP	Get correlation coefficient r. Sample value 0.976.
72	51 24	⮪, ACOS	Acos(r). Sample value 12.604° =0.220 rad.
73	25	TAN	TAN(ACOS(r))=0.224.
74	45	÷	
75	33	(	
76	22 4	RCL , 4	Get N. Sample value 4.
77	65	-	
78	2	2	
79	34	)	N-2. Sample value 2.
80	11	√x	√(N-2). Sample value 1.414.

77	65	-	
78	2	2	
79	34	)	N-2. Sample value 2.
80	11	√x	√(N-2). Sample value 1.414.
81	55	×	
82	22 3	RCL 3	Get the slope m
83	74	=	
84	26	R/S	Stop. Error in slope σ_m=0.316. Copy it and press R/S.
85	55	×	
86	33	(	
87	22 7	RCL 7	Get Σx^2=30.
88	45	÷	
89	22 4	RCL 4	Get N=4.
90	34	)	$\Sigma X^2/N$= 7.5
91	11	√x	
92	74	=	σ_b=0.866, error in intercept. End of program.

CALCULATOR: HP20S

PROCEDURE: Unweighted regression calculations with errors.

General Procedure:

1. "Fit" and error programs must first be installed, as shown above. Clear out old statistical data. Choose type of regression calculation (linear, log, exponential, power law) and, in the nonlinear case, execute (XEQ) it.
2. Enter data. Sample data, shown above for linear regression, are the same for all programs. Data entry method differs for each type of regression.
3. Run regression program or keys to find slope m and intercept b.
4. Run error program to calculate final errors.

Linear regression (unweighted, with errors)

Comments	You enter	Display, sample data
Clear statistical data.	↱, CLΣ	
Enter data.	x_1, INPUT, y_1, Σ+, ···	1.000
	x_N, INPUT, y_N, Σ+	4.000
Run error program	XEQ, 6	0.316 (σ_m)
	R/S	0.866 (σ_b)

CALCULATOR: HP20S

PROCEDURE: Exponential regression (unweighted with errors). This fits data to an exponentially varying curve $y=be^{mx}$, such as the voltage in an RC or LR circuit, or in counting experiments for radioactive decay or bacterial growth. It transforms the problem into linear regression with the equation $\ln(y)=\ln(b)+mx$, with intercept $\ln(b)$ and slope m. Results will deviate from weighted fits.

Comments	You enter	Display
Clear statistical data.	⮫, CLΣ	0.000
Start exponential program.	XEQ, B	0.000
Enter data. (Note different key R/S.)	x_1, INPUT, y_1, R/S, ···	1.000
	x_N, INPUT, y_N, R/S	4.000
Get m, b	XEQ F, ⮪, SWAP	0.653(m), 0.535(b)
Run error program	XEQ, 6	0.023 (σ_{slope})
	R/S	0.063 ($\sigma_{intercept}$)

Final answers for exponential regression for $y=be^{mx}$ from slope, intercept and errors, with examples: Values: m=0.653 b=0.535

Errors: $\sigma_m=\sigma_{slope}=0.023$ $\sigma_b=b\ \sigma_{intercept}=0.034$

Time constant=τ=1/m=1.53. Error=$\tau\ \sigma_m/m$=0.054. Cell division time=$T_{1/2}=\ln(2)/m$=1.06. Error=$T_{1/2}\ \sigma_m/m$=0.037. The same expressions hold for radioactive decay, except that (-m) is used instead of m.

CALCULATOR: HP20S

PROCEDURE: Logarithmic regression. This fits data to a logarithmic curve $y=b+m\ln(x)$, such as the potential between two cylindrical conductors. Since there is no weight factor, all calculators should give the same fit.

Comments	You enter	Display
Clear statistical data.	⮫, CLΣ	0.000
Start logarithmic program.	XEQ, C	0.000
Enter data. (Note different key R/S.)	x_1, INPUT, y_1, R/S, ···	1.000
	x_N, INPUT, y_N, R/S	4.000
Get m, b	XEQ F, ⮪, SWAP	m=4.023 b=.304
Run error program	XEQ, 6	1.262 (σ_m) 1.199 (σ_b)

CALCULATOR: HP20S

PROCEDURE: Power law regression. This fits data to a power law curve $y=bx^m$ such as the area or volume of a sphere as a function of its radius. It transforms the problem into linear regression with the equation $\ln(y)=\ln(b)+m\ln(x)$, with intercept $\ln(b)$ and slope m. Results will deviate from weighted fits.

Comments	You enter	Display
Clear statistical data.	[↱], CLΣ	0.000
Start power law program.	XEQ, A	0.000
Enter data. (Note different key R/S.)	x_1, INPUT, y_1, R/S, ···	1.000
	x_N, INPUT, y_N, R/S	4.000
Get m, b	XEQ F, [↰], SWAP	m=1.396
		b=.909
Run error program	XEQ, 6	0.159 (σ_{slope})
		0.137 ($\sigma_{intercept}$)

Final answers for power law regression for $y=bx^m$, with examples.

Values:	m=1.396	b= 0.909
Errors:	$\sigma_m=\sigma_{slope}=0.159$	$\sigma_b=b\times\sigma_{intercept}=0.137$

Trigonometric Operations on the HP-20S Calculator.*

This calculator is in the the same mode when turned on as when it was last turned off.

1. How to make conversions from one set of units to another:

   1a. Decimal degrees D.d to degrees, minutes and seconds D° M' S".

   ↱, →HMS. (See Example for the HP-20S below, line 2.)

   1b. Degrees, minutes and seconds D° M' S" to decimal degrees D.d .

   ↰ , →HR. (See Example for the HP-20S below, line 3.)

   1c. Decimal degrees D.d to radians R. ↱,→RAD. (See Example below, line 4.)

   1d. Radians R to decimal degrees D.d.

   ↰ ,→DEG (See Example for the HP-20S below, line 8.)

2. How to find the sine, cosine or tangent of an angle .

   2a. Given in degrees, minutes and seconds D° M' S". (If in radian mode, put calculator in degree mode ↱, DEG.) Convert to D.d (see 1a, Example for the HP-20S below, line 3.) Then press (SIN, COS, or TAN).

   2b. Given in decimal degrees D.d . (If in radian mode, put calculator in degree mode ↱, DEG.) Then press (SIN, COS, or TAN).

   2c. Given in radians R. (If in degree mode, put calculator in radian mode ↱ RAD, then press (SIN, COS, or TAN). (See Example, line 6.)

3. Given the sine, cosine or tangent of an angle θ, to find θ (to find $\sin^{-1}(x)$, $\cos^{-1}(x)$, or $\tan^{-1}(x)$.

   3a. In radians R. (If in degree mode, put calculator in radian mode ↱, RAD.) Enter x , press ↰ , then (ASIN, ACOS or ATAN). (See Example for the HP-20S below, line 7.)

   3b. In decimal degrees D.d. (If in radian mode, put calculator in degree mode ↱, DEG.) Enter x , press ↰ , then (ASIN, ACOS, or ATAN).

   3c. In degrees, minutes, and seconds D° M' S". (If in radian mode, put calculator in degree mode ↱, DEG.) Enter x , press ↰ , then (ASIN, ACOS, or ATAN). Convert to D.d (see 1a).

Example for the HP-20S.

Convert 1 7/8° to degrees, minutes and seconds and back again. Then convert it to radians and back again. Finally, take the sine and then get back to the original angle, both for the angle in radians and in decimal degrees.

Solution (Display fixed at 8 decimals, in degree mode).

Step	Press	Reading	Comments
1	1, +, 7, ÷, 8, =	1.87500000	1.875°, the angle in dec. deg.
2	[↱], →HMS	1.52300000	1°52'30", the angle in D°M'S"
3	[↰], →HR	1.87500000	1.875°, the angle in dec. deg.
4	[↱], →RAD	0.03272492	The angle in radians
5	[↱], RAD	0.03272492 RAD	The angle in radian mode
6	SIN	0.03271908 RAD	sin (0.03272492)
7	[↰], ASIN	0.03272492 RAD	The angle in radians
8	[↰], →DEG	1.87500000 RAD	1.875°, back to the angle in dec. deg.

*NOTE: [↱], →RAD, and [↱], RAD are two distinct operations, as are [↰], →DEG, and [↱], DEG. The operation [↱], →RAD converts the numerical value in the display from dec. deg. to radians; [↰], →DEG does the reverse. The other keys, [↱], RAD or [↱], DEG change the mode of the calculator for trigonometric operations (sine, cosine, tangent, or the inverse). The display shows "RAD" or not, depending on whether or not it is in radian mode. See this Example for using these keys.

CALCULATOR: HP48G

PROCEDURE: Entering statistical data.

(Optional) **To set display roundoff to fixed format with 4 decimals.**

1. Press ⮣, MODES to display CALCULATOR MODES.
2. Move cursor to NUMBER FORMAT.
3. Press **Choose** (B key)
4. Move cursor to **Fixed**, press **OK** (F key), then move the cursor to the number of decimals, enter the number you wish to display, for example 4, then press **OK** (F key). Display will show Fix **4**.
5. Begin data entry (see next line).

Data entry

1. Optional: To get the {HOME} screen, hold down ON, press and release C, then release ON.
2. Press ⮣, STAT. Choose single-var (mean, standard deviation, etc.) or choose **Fit data**... for two variable statistics (linear regression, exponential decay, etc.). Then press **OK** (F key).
3. Press DEL, **Reset all,** **OK** to remove old data.
4. Press **EDIT** to enter data: x_1, ENTER, ▼, x_2, ENTER, etc., until all x entries are in first column.

For two-variable data, enter data x_1, ENTER, y_1, ENTER, ▼, x_2, ENTER, y_2, ENTER, x_3, ENTER, etc., until all x- and y- entries are in first column.

When all data are in, press ENTER . Proceed to statistical calculations (see below).

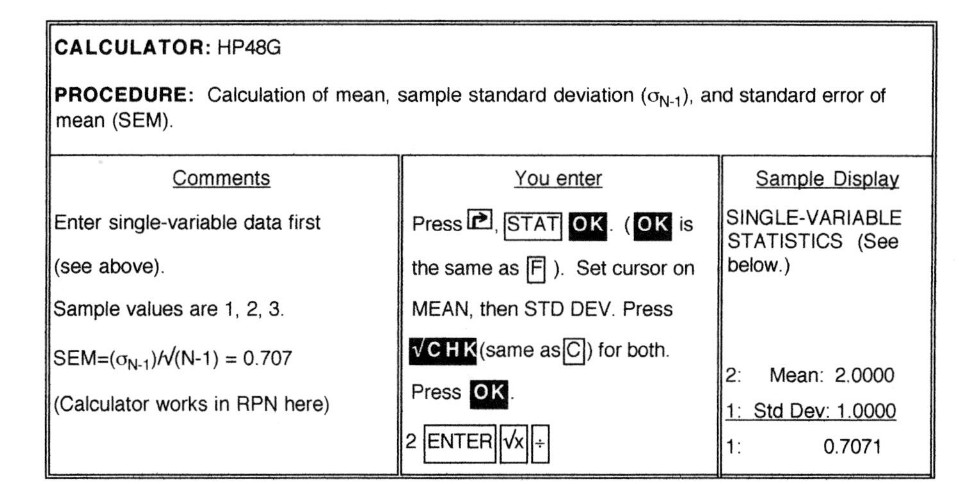

CALCULATOR: HP48G

PROCEDURE: Calculation of mean, sample standard deviation (σ_{N-1}), and standard error of mean (SEM).

Comments	You enter	Sample Display
Enter single-variable data first (see above).	Press ⮣, STAT OK. (OK is the same as F). Set cursor on	SINGLE-VARIABLE STATISTICS (See below.)
Sample values are 1, 2, 3.	MEAN, then STD DEV. Press	
SEM=(σ_{N-1})/√(N-1) = 0.707	√CHK (same as C) for both.	
(Calculator works in RPN here)	Press OK.	2: Mean: 2.0000 1: Std Dev: 1.0000
	2 ENTER √x ÷	1: 0.7071

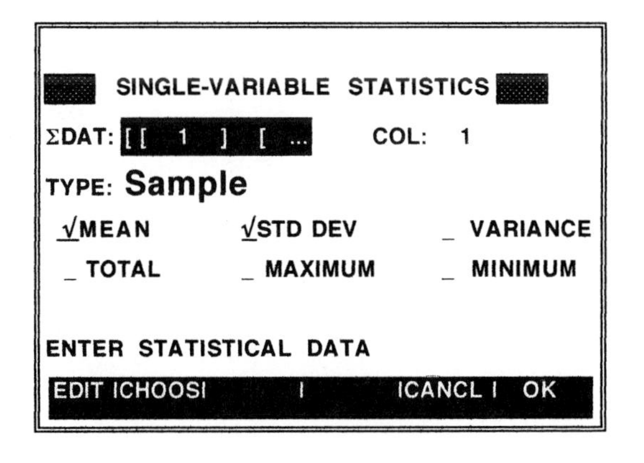

CALCULATOR: HP-48G.

PROCEDURE: Calculation of mean, sample standard deviation (σ_{N-1}), and standard error of mean (SEM) for grouped data, each with weight (frequency) y.

Sample data: x: 1 2 3 4 5

(frequency or weight) y: 4 18 24 18 4 $N=\Sigma y=68$. Mean= 3.000. $\sigma_N=1.000$.

$SEM=\sigma_N/\sqrt{(N-1)}= 0.122$

Calculation of σ_N: $\sigma_N^2 = (\Sigma yx^2)/N - (\Sigma yx/N)^2$.

The second term is $\bar{x}^2$, obtainable from data with x and y as the variables. The first term must be obtained from a second set of data with x^2 and y as the variables.

Comments	You enter	Display
Sample Data: x y x^2 1 4 1 2 18 4 3 24 9 4 18 16 5 4 25 First clear statistical registers, then enter two variable data for x and y, as shown in Data Entry section (see above). Then repeat the process for x^2 and y.	↱, STAT, move to Fit Data, OK(F), EDIT, 1, ENTER, 4, ENTER, 1, ENTER, ▼, 2, ENTER, 18, ENTER, 4, ENTER, ··· , 5, ENTER, 24, ENTER, 25, ENTER.	FIT DATA (see below. Change X-COL and Y-COL to match figure.) 1 2 3 1 1.000 4.000 1.000 2 2.000 18.0... 4.000 . . . . 5 5.000 4.000 25.0...
	ENTER, OK(F)	FIT DATA (see below)
	OK (F), DEL	{HOME} screen (see How to Get Out of Trouble below, at end of HP48G pp.).
	↰, STAT, SUMS, ΣX*Y,	2: 204.000
	ΣY,	1: 68.000
	÷	3.000 (Mean $\bar{x}$)
	↰, x^2	9.000 ($\bar{x}^2$)
	↱, STAT, move to Fit Data, OK(F)	FIT DATA (see below)
	Move to X-COL, change to 3, OK(F), ↱, STACK, move to 4, ROLL, ON, ΣX*Y, ΣY, ÷	10.000 $(\Sigma yx^2)/N$
	↰, SWAP, -	1.000 (σ_N)
	ΣY, 1, -, √x, ÷	0.122 (SEM)

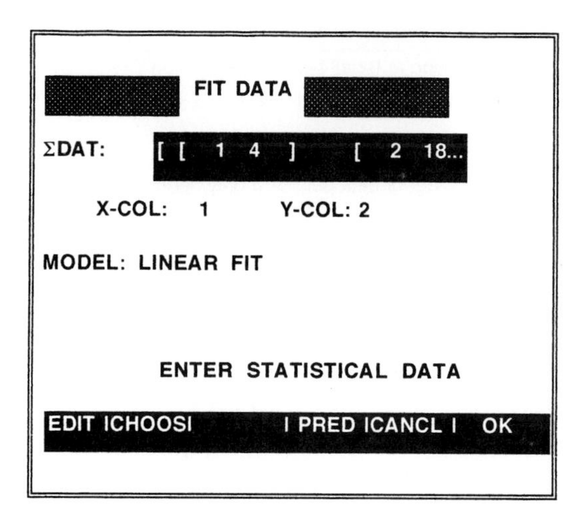
FIT DATA
ΣDAT: [[1 4] [2 18...
X-COL: 1 Y-COL: 2
MODEL: LINEAR FIT
ENTER STATISTICAL DATA
EDIT ICHOOSI I PRED ICANCL I OK

Program Title: LRE- Linear Regression With Errors **Calculator:** HP48G	
Description: This program finds the errors in the slope m for all types of regression calculations and in the intercept b for linear fits only. It uses the algorithms $\sigma_m = \lvert m\rvert\sqrt{\frac{\left(\frac{1}{r}\right)^2 - 1}{n-2}}$, $\sigma_b = \sigma_m\sqrt{\frac{\Sigma x^2}{n}}$, where n is the number of data points. It only needs to be keyed in once and stored permanently. It is then ready to calculate errors. First, the Linear Regression Calculation, which follows this page, should be completed to the point marked "End of calculation without errors. " Then this program can be run at the point marked "Error calculation." The display is for a line fit (LINFIT) to the following sample data: x 1 2 3 4 intercept b= -1.5, slope m=2.0 y 1 2 4 7 errors: σ_b=0.866 σ_m=0.316.	
Key	**Display. Comments**
	Before running this program, display should appear as shown. 3: '-1.500+2*X' 2: Correlation: 0.976 1: Covariance: 3.333
DEL, ↰, STAT, **FIT**(E), **LR**(B)	Intercept: -1.500 Slope: 2.000
MATH, **REAL**, NXT, **ABS**(A), ↰, STAT, **FIT**(E)	These steps, which avoid a negative error, are optional. Without them , simply ignore the negative sign which goes with a negative slope.
CORR, 1/x, ↰, x^2, 1, -, ↰, STAT, **SUMS**, **NΣ** (F), 2, -, ÷, √x, ×, ↰, "", α, α, ERRM, α, ENTER, ↰, SWAP	3: Intercept: -1.500 2: "ERRM=" 1: 0.316 (σ_m, the error in the slope)
↰, STAT, **SUMS** (F), **Σx^2**(C), **NΣ**(F), ÷, √x, ×	3: Intercept: -1.500 2: "LINFIT ONLY:ERRB=" 1: 0.866 (σ_b, the error in the slope) Optional program to calculate error in the intercept. This is only applicable to LINFIT, linear correlation. It does not give the correct result for logarithmic, exponential, nor power law fits.

Listing of the program "LRE":	
⮫, MEMORY, move	
to LRE, EDIT, EDIT,	<<LR ABS CORR INV
	SQ 1 - NΣ 2 - / √ *
	"ERRM" SWAP DUP (Calculation of σ_m)
	ΣX^2 NΣ / √ *
▼, ▼, ▼, ▼	"LINFIT ONLY:ERRB=" (Calculation of σ_b)

CALCULATOR: HP48G

PROGRAM: Linear Regression Calculations

First, enter two columns of data, x and y, in the calculator. Then decide on the form of regression calculation: linear, logarithmic, exponential, or power law fit. The calculator converts the problem into a linear regression calculation of the following form:

Linear fit (Linfit)	$y=b+mx$,
Logarithmic fit	$y=b+m \ln x$,
Exponential fit	$y=be^{mx}$, or $\ln y=\ln b + mx$,
Power law fit	$y=bx^{m}$, or $\ln y=\ln b + m \ln x$.

You next use the program "LRE," which you already have keyed into the calculator (it remains stored in the memory), to find the error in the slope (difference quotient) *m* for all types of fits. The program calculates the error in the intercept *b* correctly for Linfit only. (One seldom needs the error in *b*.) All error calculations are based on data scatter, use unweighted data, and thus will give results differing from the more accurate weighted fits.

Sample calculation is a line fit. Data: x 1 2 3 4 intercept b= -1.5, slope m=2.0

y 1 2 4 7 errors: σ_b=0.866 σ_m=0.316.

Comments	You enter	Sample Display
Choose linear regression	[↱], STAT, move to Fit Data, **OK** (F)	FIT DATA (see above)
Enter data (see "entering statistical data")	1, [ENTER], 1, [ENTER], [▼], 2, [ENTER], 2, [ENTER], 3, [ENTER], 4, [ENTER], 4, [ENTER], 7, [ENTER]	1 2 1 1.000 1.000 2 2.000 2.000 3 3.000 4.000 4 4.000 7.000
Accept data, move to FIT DATA menu. check column locations (x-1, y-2), move to model and choose fit , if necessary.	[ENTER], **CHOOS** (B), **Linear Fit**	FIT DATA (see above) ΣDAT **[[1 1] [2 2 ...** X-COL 1 Y-COL 2 MODEL: **Linear Fit**
Accept data, choice of fit, and get equation for fit. End of calculation without errors.	**OK**	3: '-1.500+2*X' 2: Correlation: 0.976 1: Covariance: 3.333
Error calculation. First, program "LRE" must have been entered and the calculations above should have been completed. Note: ERRB is only correct for linear fits.	DEL, VAR, **LRE** (press letter immediately underneath)	4: "ERRM=" 3: 0.316 2: "LINFIT ONLY:ERRB=" 1: 0.866

Trigonometric Operations on the HP 48G Calculator.

This calculator is in the radian mode if it reads "RAD" in the upper left hand corner. It is in the decimal degrees mode if it reads nothing. To toggle from one mode to the other, press ↰, RAD. (To get out of grad mode, press ↱, MODES, move to ANGLE MEASURE and press +/-, OK.)

1. How to make conversions from one set of units to another:

   1a. Decimal degrees D.d to degrees, minutes, and seconds D° M' S".
   D.d, ENTER, ↰, TIME, NXT, ->HMS .(See Example below, step 2.)

   1b. Degrees, minutes and seconds D° M' S" to decimal degrees D.d.
   D. MMSS, ENTER, ↰, TIME, NXT, HMS-> . (See Example below, step 3.)

   1c. Decimal degrees D.d to radians R.
   D.d, ENTER, MTH, REAL , NXT, NXT, D->R (See Example below, step 4.)

   1d. Radians R to decimal degrees D.d. R, ENTER, MTH, REAL , NXT, NXT, R->D . (See Example below, step 5.)

2. How to find the sine, cosine, or tangent of an angle

   2a. Given in degrees, minutes, and seconds D° M' S". Follow 1b, (sin, cos, or tan).

   2b. Given in decimal degrees D.d. D.d, (sin, cos, or tan). (See Example below, step 6.)

   2c. Given in radians R. Put in Rad mode, R, (sin, cos or tan.)

3. Given x, the sine, cosine or tangent of an angle θ, to find θ [to find $\sin^{-1}(x)$, $\cos^{-1}(x)$, or $\tan^{-1}(x)$].

   3a. In radians R. Put in Rad mode, enter value of ($\sin\theta$, $\cos\theta$ or $\tan\theta$), ↰, (ASIN. ACOS, or ATAN). (See Example below, step 7.)

   3b. In decimal degrees D.d. Put in degree mode, enter value of ($\sin\theta$, $\cos\theta$, or $\tan\theta$), ↰, (ASIN. ACOS, or ATAN).

   3c. In degrees, minutes, and seconds D° M' S". In degree mode, follow step 1b, then ↰, (ASIN. ACOS, or ATAN).

Example for the HP 48G

Convert 1 7/8° to degrees, minutes and seconds and back again. Then convert it to radians and back again. Find its sine and use the inverse sine to find its value in radians.

Solution (In degree mode, fixed at 8 decimals)

Step	Press	Reading	Comments
1	1, ENTER, 7, ENTER, 8, ENTER, ÷, +	1.87500000	1.875°, the angle in dec. degr.
2	↰, TIME, NXT, ->HMS .	1.52300000	1°52'30", the angle in D.MMSS

3	HMS ->	1.87500000	1.875°, the angle in dec. degr.
4	MTH, REAL, NXT, NXT,		
	D->R	0.03272492	The angle in radians
6	R->D	1.87500000	1.875°, the angle in dec. degr.
7	Sin	0.03271908	sin(1.875°)
8	↰, RAD, ↰, ASIN	0.03272492	The angle in radians

How to Get Out of Trouble with the HP48G

1. If you are entering data or are setting up an operation, and nothing happens, try pressing ENTER Or press OK (F6).
2. To start over again, press CANCEL several times, until you see the normal display:

```
{ home }              (date)    (time)
4:
3:
2:
1:
VECTRI MATR I LIST I HYP I REAL I BASE
```

3. If that doesn't work and you see a " ⧗ " in the annunciator area (at the top), press and hold ON, press and release the "C" menu key (in top row), then release ON. This will destroy data on the "stack" (data area on the normal display).
4. As a last resort, press and hold ON, press and release the left and right menu keys (A and F on the top row), release ON, and press NO. Warning: This act of desperation will destroy all your memory (programs and data)!

For other emergency measures, see "HP 48G Series Quick Start Guide", pp. 9-2 to 9-4.

CALCULATOR: SHARP EL-520L

PROCEDURES: All.

Use the programs for the Sharp EL-506L, EL-546L, with the following changes:

Data Entry. For single-variable statistical calculations (SD) press MODE, 1 (stat x) , for two-variable (regression) calculations, press MODE, 2 (stat xy).

Program Title: Linear Regression (continued)

Instead of MODE, 3 (STAT) 1(a+bx), substitute MODE 2 (stat xy).

Procedure: Logarithmic regression.

Instead of MODE, 3 (STAT), 4(lnx), substitute MODE 2 (stat xy).

Instead of x_1, (x,y), y_1, DATA, substitute ln, x_1, (x,y), y_1, DATA. Do same for x_2, x_N, etc.

Procedure: Exponential regression.

Instead of MODE, 3 (STAT), 3(e^x), substitute MODE 2 (stat xy).

Instead of x_1, (x,y), y_1, (x,y), y_1^2, DATA, substitute x_1, (x,y), ln, y_1, (x,y), y_1^2, DATA. Same for x_2, x_N, etc.

Instead of RCL, a, substitute 2ndF, e^x, RCL, a.

Procedure: Exponential regression (counts, weighted, with errors).

Instead of MODE, 3 (STAT), 3(e^x), substitute MODE 2 (stat xy).

Instead of x_1, (x,y), y_1, (x,y), y_1, DATA, substitute x_1, (x,y), ln, y_1, (x,y), y_1, DATA. Same for x_2, x_N, etc.

Instead of RCL, a, substitute 2ndF, e^x, RCL, a.

Procedure: Power law regression.

Instead of MODE, 3 (STAT), 5(ax^b), substitute MODE 2 (stat xy).

Instead of x_1, (x,y), y_1, (x,y), y_1^2, DATA, substitute ln, x_1, (x,y), ln, y_1, (x,y), y_1^2, DATA. Same for x_2, x_N, etc.

Instead of RCL, a, substitute 2ndF, e^x, RCL, a.

CALCULATOR: SHARP EL-506L, EL-546L

PROCEDURE: Entering statistical data.

Set-up for display roundoff. (Permanent: do only on first use.)

Press 2ndF, FSE and repeat until FIX (or SCI) is shown in the display. Then press 2ndF, TAB, 3, if you wish to display 3 decimals. It is advisable to set the display to 8 or more decimals to avoid round-off errors, which can occur during calculation of scientific functions of memory contents.

Data entry

For all statistical calculations, press MODE, 3 (STAT). Then proceed as instructed below.

1. Single variable (mean, standard deviation and standard error of the mean for N entries, one or more at each value of x).
   a. Press 0 (SD) to prepare for single variable data entry.
   b. (Optional) MODE, 3 automatically will remove old data. To remove all old data, also press 2ndF, CA. To check for clearance, press RCL, n. Display should read n=0.000.
   c. Enter data, for one entry at each value of x: press x_1, DATA, x_2, DATA,$\cdots x_N$, DATA .
2. Single variable (with a frequency or weight f_i at x_i) is the same as part 1, except for
   c. Enter data: press x_1, (x,y), f_1, DATA, x_2, (x,y), f_2, DATA,$\cdots x_N$, (x,y), f_N, DATA .
3. Two-variables (regression calculations)
   a. Press MODE, 3 (STAT), 1 (a+bx) to prepare for two variable data entry.
   b. (Optional) To clear old data, press 2ndF, CA. To check for clearance, press 2ndF, n. Display should read 0.000 (0.00000000 or 0.00000000×10^{00} for regression calculations with errors. See "Setup for display roundoff" above.)
   c. For unweighted, single y-value for each x-value, linear regression without error calculation, enter data: press x_1, (x,y), f_1, DATA, x_2, (x,y), f_2, DATA,$\cdots x_N$, (x,y), f_N, DATA .
   d. See the bracketed programs [linear, logarithmic, exponential, exponential (counts), and power law regression, all with error computation and/or with weights] for special forms of input.

CALCULATOR: SHARP EL-506L, EL-546L

PROCEDURE: Calculation of mean, sample standard deviation (σ_{N-1}), standard error of mean (SEM).

Comments	You enter	Display
First clear statistical registers, then enter single-variable data, as shown in Data Entry section. (See immediately above.) Sample values are 1, 2, 3. σ_{N-1}=1.0, σ_N=0.816 SEM=(σ_{N-1})/√(N) = 0.577	1. Press 3, DATA	n=1.000
	2. Press 2, DATA·	n=2.000
	3. Press 1, DATA·	n=3.000
	4. Press RCL, $\overline{x}$	2.000 (mean)
	5. Press RCL, s_x	1.0 (σ_{N-1})
	6. Press ÷, √, 3, =	.577 (SEM)

CALCULATOR: SHARP EL-506L, EL-546L.

PROCEDURE: Calculation of mean, sample standard deviation (σ_{N-1}), standard error of mean (SEM) for grouped data, each with weight (frequency) f.

Sample data: x: 1 2 3 4 5

(frequency or weight) f: 4 18 24 18 4 N=68. Mean= 3.000. σ_N =1.000.

SEM=σ_{N-1}/√(N)= 0.122

Comments	You enter	Display
First clear statistical registers, then enter one variable data for x and f, as shown in Data Entry section 2 (see above).	5, (x,y), 4, DATA·	n=4.000
	4, (x,y), 18, DATA·	n=22.000
	. . . . . .	. . . . . .
	1, (x,y), 4, DATA·	n=68
	RCL, $\overline{x}$	$\overline{x}$=3.000
	RCL, s_x	σ_{N-1}=s_x=1.007
	÷, √, RCL, n	68.000 (N=∑ f)
	=	0.122 (SEM)

CALCULATOR: SHARP EL-506L, EL-546L.

PROCEDURE: Linear Regression (unweighted, without errors) uses the built-in calculation.

Comments	You enter	Display
First clear statistical registers, then enter two-variable data for x and y, as shown in Data Entry section (see above).	1, (x,y), 1, DATA·	n=1.000
	2, (x,y), 2, DATA·	n=2.000
	3, (x,y), 4, DATA·	n=3.000
Sample data:	4, (x,y), 7, DATA·	n=4.000
x: 1 2 3 4	RCL, b	2.000. b.
y: 1 2 4 7	RCL, a	-1.500. a.
y=a+bx	RCL, r	0.976. r.
intercept a=-1.500		
slope b=2.000		
correlation coefficient r=0.976		

CALCULATOR: SHARP EL-506L, EL-546L.

PROCEDURE: Linear Regression. Programs to enter weighted data and to find errors in slope b and intercept a. These programs can be used for all four types of regression (linear, log, exponential, and power law), which differ slightly in data entry and treatment of errors. Because these programs use weighted data, results will differ from less accurate programs which use unweighted data. Since these calculators lack programmability (i.e. can't learn a program), you must key in all the steps each time you do a calculation. Each program reduces the problem to a linear regression fit of the form y=a+bx and builds on the standard, built-in algorithms for a and b. Error algorithms are based on data scatter:

$$\sigma_b = b\sqrt{\frac{\frac{1}{r^2} - 1}{N-2}} \quad \text{and} \quad \sigma_a = \sigma_b\sqrt{\frac{\Sigma x^2}{n}}$$

Note that N is the number of x-values, which you will put in by hand, and n is the sum of the weights, which the program will insert automatically. You first put in the data with the appropriate weights, if necessary, then run the built in linear regression calculation to get the slope and intercept, run the error computation program (by hand) and then calculate the final errors.

Sample data are the same for all programs:

x: 1 2 3 4

y: 1 2 4 7

<u>To avoid ambiguity in all the programs, a and b refer to intercept and slope on row five of the keyboard; C, D, E, and F refer to storage registers on row 3.</u>

Program Title: Linear Regression (continued). Program for straight line, linear regression, y=a+bx, to enter unweighted data and to find errors.

Calculator: SHARP EL-506L, EL-546L.

Sample values are x, y: 1,1 2,2 3,4 4,7. b=2, a=-1.5, σ_b=0.31622779, σ_a=0.86602546. See above for error algorithms.

Press	Display, Sample Value, Comments
2ndF, FSE, (repeat if necessary), 2ndF, TAB, 8	FIX (or SCI), 0.00000000 (or $0.00000000 \times 10^{00}$). Set calculator to 8 or more decimals to avoid round-off errors.
MODE, 3 (STAT), 1(a+bx), 2ndF, CA, RCL, n	n=0.00000000. Put calculator in linear regression mode, check for clearance of stat registers.
x_1, (x,y), y_1, DATA·	n=1.00000000.
x_2, (x,y), y_2, DATA·	n=2.00000000.
. . . . .	. . . . . .
x_N, (x,y), y_N, DATA·	n=4.00000000.
RCL, a (button on row 5)	-1.50000000. Intercept a. Write it down.
RCL, b (button on row 5)	2.00000000. Slope b. Write it down.
×, √, (, (, RCL, r, 2ndF, x^{-1}, x^2, -, 1,), ÷, (, N, - , 2,),), =	0.31622779. σ_b. N=number of data points (4 here).
×, √, (, RCL, Σx^2, ÷, RCL, n,), =	0.86602547. σ_a

CALCULATOR: SHARP EL-506L, EL-546L

PROCEDURE: Logarithmic regression. This fits data to a logarithmic curve y=a+bln(x), such as the potential between two cylindrical conductors. Sample values are x, y: 1,1 2,2 3,4 4,7. b=4.02286063, a=0.30378309, σ_B=1.26205757, σ_A=1.19882417. See above for error algorithms.

Press	Display, Sample Value, Comments
2ndF, FSE, (repeat if necessary), 2ndF, TAB, 8	FIX (or SCI), 0.00000000 (or $0.00000000 \times 10^{00}$). Set calculator to 8 or more decimals to avoid round-off errors.
MODE, 3 (STAT), 4(ln x), 2ndF, CA, RCL, n	0.00000000. Put calculator in logarithmic regression mode, check for clearance of statistical registers.
x_1, (x,y), y_1, DATA·	n=1.00000000.
x_2, (x,y), y_2, DATA·	n=2.00000000.
.....	..... .
x_N, (x,y), y_N, DATA·	n=4.00000000.
RCL, a (button on row 5)	0.30378309. Intercept a. Write it down.
RCL, b (button on row 5)	4.02286063. Slope b. Write it down.
×, √, (, (, RCL, r, 2ndF, x^{-1}, x^2, -, 1,), ÷, (, N, - , 2,),), =	1.26205757. σ_b. N=number of data points (4 here).
×, √, (, RCL, Σx^2, ÷, RCL, n,), =	1.19882417. σ_a

CALCULATOR: SHARP EL-506L, EL-546L

PROCEDURE: Exponential regression (weighted, with errors). This fits data to an exponentially varying curve $y=ae^{bx}=e^{A+bx}$, such as the voltage in an RC or LR circuit. It uses weights=y^2 to transform the problem into linear regression with the equation ln(y)=ln(a)+bx, with intercept A=ln(a) and slope b. Sample values are x, y, w: 1,1,1; 2, 2, 4; 3, 4, 16; 4, 7, 49; $a=e^A=0.61361307$, $b=0.61063339$, $\sigma_b=0.02967467$, $\sigma_A=0.10903169$, $\sigma_a=a\sigma_A=0.06690327$. Unweighted fits will deviate from these results.

Press	Display, Sample Value, Comments
2ndF, FSE, (repeat if necessary), 2ndF, TAB, 8	FIX (or SCI), 0.00000000 (or 0.00000000×10^{00}). Set calculator to 8 or more decimals to avoid round-off errors.
MODE, 3 (STAT), 3 (e^x), 2ndF, CA, RCL, n	0.00000000. Put calculator in exponential regression mode, check for clearance of statistical registers.
x_1, (x,y), y_1, (x,y), y_1^2, DATA·	n=1.00000000.
x_2, (x,y), y_2, (x,y), y_2^2, DATA·	n=5.00000000.
.....	..... .
x_N, (x,y), y_N, (x,y), y_N^2, DATA·	n=70.00000000.
RCL, a (button on row 5)	0.61361307. a. Write it down.
RCL, b (button on row 5)	0.61063339. b. Write it down.
×, √, (, (, RCL, r, 2ndF, x^{-1}, x^2, -, 1,), ÷, (, N, -, 2,),), =	0.02967467. σ_b. N=number of data points (4 here). Note that N≠n. $n=\sum_{i=1}^{N} f_i$
×, √, (, RCL, Σx^2, ÷, RCL, n,), =	0.10903169. σ_A
×, RCL, a, =	0.06690327. $\sigma_a = a\,\sigma_A$

Time constant $\tau=1/B=1.64$. Error=$\tau\sigma_B/B=0.08$. . Use the same expressions for decaying exponentials, with -B instead of B.

CALCULATOR: SHARP EL-506L, EL-546L

PROCEDURE: Exponential regression (counts,weighted, with errors). This fits data to an exponentially varying curve $y=ae^{bx}=e^{A+bx}$, such as counting experiments for radioactive decay or bacterial growth. It uses weights=y to transform the problem into linear regression with the equation ln(y)=ln(a)+bx, with intercept A=ln(a) and slope b. Sample values are x, y, w: 1,1,1; 2, 2, 2; 3, 4, 4; 4, 7, 7; b=0.634, σ_B=0.0.02755003, σ_A=0.09225883, a=e^A=0.56616027, σ_a= $a\sigma_A$=0.05223328. Unweighted fits will deviate from these results.

Press	Display, Sample Value, Comments
2ndF, FSE, (repeat if necessary), 2ndF, TAB, 8	FIX (or SCI), 0.00000000 (or 0.00000000×10⁰⁰). Set calculator to 8 or more decimals to avoid round-off errors.
MODE, 3 (STAT), 3 (e^x), 2ndF, CA, RCL, n	0.00000000. Put calculator in exponential regression mode, check for clearance of statistical registers.
x_1, (x,y), y_1, (x,y), y_1, DATA·	n=1.00000000.
x_2, (x,y), y_2, (x,y), y_2, DATA·	n=3.00000000.
.....	..... .
x_N, (x,y), y_N, (x,y), y_N, DATA·	n=14.00000000.
RCL, a (button on row 5)	0.56616027. a. Write it down.
RCL, b (button on row 5)	0.63371413. b. Write it down.
×, √, (, (, RCL, r, 2ndF, x^{-1}, x^2, -, 1,), ÷, (, N, -, 2,),), =	0.02755003. σ_b. N=number of data points (4 here). Note that N≠n. $n=\sum_{i=1}^{N} f_i$
×, √, (, RCL, $\sum x^2$, ÷, RCL, n,), =	0.09225883. σ_A
×, RCL, a, =	0.05223328. $\sigma_a = a\,\sigma_A$

Division time $T_{1/2}$=ln(2)/B=1.09 Error= $T_{1/2}\sigma_B$/B=0.05. Use the same expressions for radioactive decay, with -B instead of B.

CALCULATOR: SHARP EL-506L, EL-546L

PROCEDURE: Power law regression. This fits data to a power law curve $y=ax^b$ such as the area or volume of a sphere as a function of its radius. It uses weights=y^2 to transform the problem into linear regression with the equation ln(y)=ln(a)+bx, with intercept A=ln(a) and slope b. Sample values are x, y, w: 1,1,1; 2, 2, 4; 3, 4, 16; 4, 7, 49; b=1.65933845, σ_b=0.17490288, rounded to 0.17. σ_A=0.22457127, a=e^A=0.68796191, $\sigma_a = a\sigma_A$=0.15. Unweighted fits will deviate from these results.

Press	Display, Sample Value, Comments
2ndF, FSE, (repeat if necessary), 2ndF, TAB, 8	FIX (or SCI), 0.00000000 (or 0.00000000×10^{00}). Set calculator to 8 or more decimals to avoid round-off errors.
MODE, 3 (STAT), 5 (ax^b), 2ndF, CA, RCL, n	0.00000000. Put calculator in power law regression mode, check for clearance of statistical registers.
x_1, (x,y), y_1, (x,y), y_1^2, DATA·	n=1.00000000.
x_2, (x,y), y_2, (x,y), y_2^2, DATA·	n=5.00000000.
. . . . .	. . . . .
x_N, (x,y), y_N, (x,y), y_N^2, DATA·	n=70.00000000.
RCL, a (button on row 5)	0.68796191. a. Write it down.
RCL, b (button on row 5)	1.65933845. b. Write it down.
×, √, (, (, RCL, r, 2ndF, x^{-1}, x^2, -, 1,), ÷, (, N, -, 2,),), =	0.17490288. σ_b. N=number of data points (4 here). Note that N≠n. $n = \sum_{i=1}^{N} f_i$
×, √, (, RCL, Σx^2, ÷, RCL, n,), =	0.22457127. σ_A
×, a, =	0.15. $\sigma_a = a\sigma_A$

Trigonometric Operations on Sharp Calculators.

This calculator's mode (DEG, RAD, GRD) is shown on the display. To change it, press DRG until you get the desired mode. Press MODE 0 to get into normal mode.

1. How to make conversions from one set of units to another:

1a. Decimal degrees D.d to degrees, minutes, and seconds D° M' S" .

2ND F, <->DEG. (See Example below, line 2.)

1b. Degrees, minutes and seconds D° M' S" to decimal degrees D.d .

To enter an angle given in D° M' S", press D, D° M' S, M, D° M' S, S, D° M' S. Angle in D° M' S" will be in display. To convert D° M' S" to D.d, press 2ND F, <->DEG. (See Example below, line 3.)

1c. Decimal degrees D.d to radians R.

Press DRG until you are in DEG mode. Enter the angle in D.d, press 2nd F, DRG-> until display reads ANS->RAD. (See Example, lines 4,5.)

1d. Radians R to decimal degrees D.d .

Press DRG until you are in RAD mode. Enter the angle in radians, press 2nd F, DRG-> until display reads ANS->DEG. (See Example, line 6.)

2. How to find the sine, cosine, or tangent of an angle θ.

2a. given in degrees, minutes and seconds D° M' S" or in decimal degrees.

Press DRG until you are in DEG mode. Press SIN, θ (enter either as D.d, or as D° M' S"-see 1b), =.

2b. Given in radians R.

Press DRG until you are in RAD mode. Press SIN, θ (in radians), =.

3. Given the sine, cosine or tangent of an angle θ, to find θ [to find $\sin^{-1}(x)$, $\cos^{-1}(x)$, or $\tan^{-1}(x)$].

3a. in radians R.

Press DRG until you are in RAD mode, then 2ND F, SIN^{-1} (or COS^{-1} or TAN^{-1}), x, =. (See Example, line 5.)

3b. in decimal degrees D.d .

Press DRG until you are in DEG mode, then 2ND F, SIN^{-1} (or COS^{-1} or TAN^{-1}), x, =. (See Example, line 4.)

3c. in degrees, minutes, and seconds D ° M' S". Follow 3b, but enter x in D ° M' S" as in 1b.

Example for Sharp Calculators (with calculator fixed to 8 decimals, in DEG mode and normal modes). Convert 1 7/8° to degrees, minutes, and seconds and back again. Take the sine and then get back to the original angle. Convert the angle to radians and back to decimal degrees.

Solution.

Step	Press	Reading	Comments
1	1, +, 7, ÷, 8, =	1.87500000	1.875°, the angle in dec. deg.
2	2ND F,<->DEG	1°52°30.00	1°52'30", the angle in D° M' S"
3	2ND F,<->DEG	1.87500000	1.875°, the angle in dec. deg.
4	STO A, SIN, RCL A, =	0.03271908	sin(1.875°)
5	STO B, 2ND F, SIN^{-1}, =	1.87499984	1.875°, the angle in dec. deg.
6	press 2nd F, DRG->	ANS->RAD	
		0.03272492	The angle in radians
7	2nd F, DRG->,	ANS->DEG	
	2nd F, DRG->	1.87499984	1.875°, the angle in dec. deg.

Common Problems with TI Graphing Calculators	
Problem	**Solution**
Displays wrong key code (alphabetical *vs.* non-alphabetical)	Press **ALPHA** (toggles alphabetical codes on or off).
(TI-86 only) Can't get to bottom line on screen	Press the F-key under the item you want.
Calculator won't enter my command or won't take me where I want to go.	Press **ENTER**
I can't escape from where I am.	Press **2nd** **QUIT**, **EXIT** or **CLEAR**. (Warning: **CLEAR** erases the line of the program you are editing.)

CALCULATOR: TI-80 (Also for TI-82, 83, 86 with minor changes)

PROCEDURE: Entering statistical data.

Setup for display roundoff and list names. (Permanent: do only on first use.)

1. Press MODE to display the MODE screen. Leave it on NORMAL.
2. Move cursor to second line, then move it to the right to the number of decimals you wish to display, for example 3, and press ENTER, then 2nd QUIT.
3. Press STAT (or 2nd STAT) EDIT. Name columns 1, 2, 3, (number in upper right hand corner) L1, L2, L3. Move cursor to top of column and type in correct name and press ENTER. (Press ALPHA to toggle between alphabetical and nonalphabetical key codes. To change a name, use DEL & 2nd INS)
4. Begin data entry (see next line).

Data entry

1. Press STAT (or 2nd STAT) EDIT. (optional) To remove old data, go to top of column, press CLEAR ENTER.
2. Enter data x_1, ENTER, x_2, ENTER, etc., until all x entries are in first column. Repeat in second column for y-entries for two-variable data. If necessary, enter further data in third column for frequency if needed. (See below.) QUIT or EXIT, and proceed to statistical calculations (see below).

CALCULATOR: TI-80 (Also for TI-82, 83, 86 with minor changes)

PROCEDURE: Calculation of mean, sample standard deviation (σ_{N-1}), standard error of mean (SEM).

Comments	You enter	Display
Enter single variable data (see above). Then quit or exit edit menu.	1. Press [STAT], go to CALC	EDIT CALC
		1: 1-VAR STATS
	Press [ENTER] (or F1)	1-VAR STATS (or OneVar)
Sample values are 1, 2, 3.	2. [2nd] [L1] (or [ALPHA] L 1) [ENTER]	x=2.000
		ΣX=6.000
		(ΣX²=14.000)
		SX=1.000
		σX=.816
		n=3.0000
SEM=(σ_{N-1})/√(N) = 0.577	3. 1 [÷] [2nd] [√] 3 [ENTER]	1/√3 (or 1/√(3)
		.577

CALCULATOR: TI-80 (Also for TI-82, 83, 86 with minor changes)

PROCEDURE: Calculation of mean, sample standard deviation (σ_{N-1}), standard error of mean (SEM) for grouped data.

Comments	You enter	Display
Enter single variable data, then quit or exit edit menu.	1. Press STAT , go to CALC	EDIT CALC
Sample values are shown as entered	Press ENTER (or F1)	1:1-VAR STATS (or One Var)
L1(x) L2(frequency)		1-VAR STATS
1 4		x=3.000
2 18	2. 2nd L1 , 2nd L2	ΣX=204.000
3 24	(or ALPHA L 1 , ALPHA	(ΣX²=680.000)
4 18	L ALPHA 2) ENTER	SX=1.007
5 4		σX=1.000
SEM=(σ_{N-1})/√(N) = 0.122		MED=3.000
		n=68.000
	3. 1.007 ÷ 2nd √ 68	1.007/ √68 (or 1.007/ √(68)
	ENTER	.122

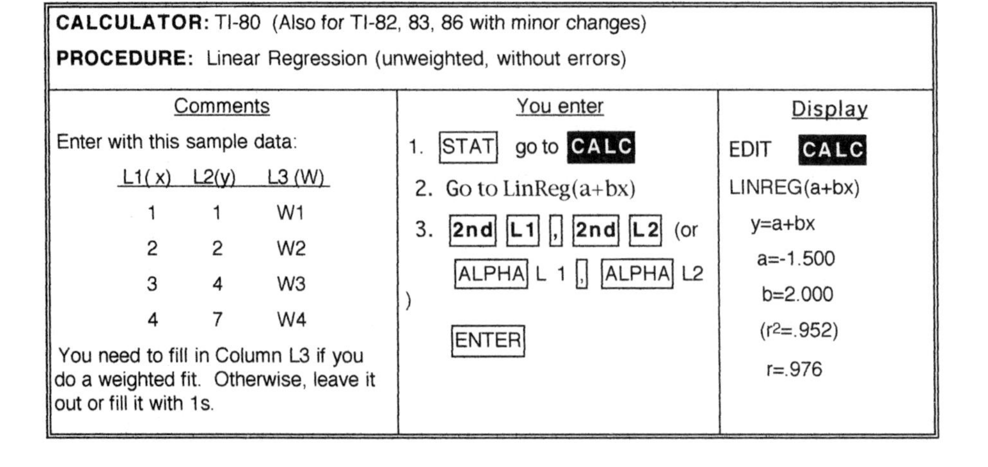

CALCULATOR: TI-80 (Also for TI-82, 83, 86 with minor changes)

PROCEDURE: Linear Regression (unweighted, without errors)

Comments	You enter	Display
Enter with this sample data:	1. STAT go to CALC	EDIT CALC
L1(x) L2(y) L3 (W)	2. Go to LinReg(a+bx)	LINREG(a+bx)
1 1 W1	3. 2nd L1 , 2nd L2 (or	y=a+bx
2 2 W2	ALPHA L 1 , ALPHA L2	a=-1.500
3 4 W3	)	b=2.000
4 7 W4	ENTER	(r²=.952)
You need to fill in Column L3 if you do a weighted fit. Otherwise, leave it out or fill it with 1s.		r=.976

CALCULATOR: TI-80 (Also for TI-82, 83 with minor changes. See separate program for TI-86.))

PROCEDURE: Linear Regression. Program to find errors in a and b. This program, with small additions, can be used for all four types of regression (linear, log, exponential, and power law).

Comments	You enter	Display
Before you can find the errors you must type in the data, this program, and do the regression calculation.	1. PRGM go to NEW	PROGRAM
Name the prgm "LRE".	2. LRE ENTER	NAME=
	3. VARS go to STATISTICS	PROGRAM:NEW
		PROGRAM:LRE
Enter the program.	go to EQ go to b ENTER 2nd	:b√((r-12-1)/(DIM
WARNING: Use 2nd QUIT to	√((VARS go to STATISTICS	L1-2))->E
get out of editing. The CLEAR button will erase a whole line of the program.	go to EQ go to r ENTER	:DISP "ERRB=",E
NOTE: Some calculators automatically enter "√ (" when you type "√ ".	x-1 x2 - 1) ÷ (2nd LIST	:DISP "ERRA=",E√
	go to DIM 2nd L1 - 2))	(∑X2/n)
	STO▶ ALPHA E ENTER	:
	PRGM go to I/O go to DISP	
	ENTER 2nd A-LOCK " E	
	R R B 2nd TEST ENTER	
	ALPHA " , ALPHA E	
	ENTER PRGM go to I/O go	
	to DISP ENTER 2nd	
	A-LOCK " E R R A 2nd	
	TEST ENTER ALPHA	
	" , ALPHA E 2nd √ (	
	VARS go to STATISTICS	
	go to Σ go to Σx2 ENTER	
	÷ VARS go to STATISTICS	
	go to n ENTER) 2nd QUIT	

CALCULATOR: TI-86

PROCEDURE: Linear Regression. Program to find errors in a and b. This program, with small additions, can be used for all four types of regression (linear, log, exponential, and power law).

Comments	You enter	Display
Before you can find the errors you must type in the data, this program, and do the regression calculation. Name the prgm "LRE". Enter the program. WARNING: Use EXIT to get out of editing. The CLEAR button will erase a whole line of the program.	1. PRGM go to EDIT (F2)	PROGRAM
	2. LRE ENTER	NAME=
	3. 2nd CATLG-VARS go to ALL (F2) go to b ENTER 2nd √((2nd CATLG-VARS go to ALL (F2) go to corr ENTER 2nd x-1 x2 - 1) ÷ (2nd LIST go to OPS(F4) go to dimL (F1) ALPHA L,1 - 2)) STO► E ENTER	PROGRAM:LRE :b√((corr-12-1)/(dimL L1-2))->E
Note on TI-86. Instead of using L1, L2, L3, for column titles and statistics commands, the TI-86 has "xStat", "yStat", and "fStat", which are automatic (You can leave out the "L1", "L2", "L3" in commands.) If you use these, you must enter "fStat", instead of "L1" in the LRE program (i.e. DimL 2nd ALPHA F ALPHA S 2nd ALPHA T 2nd ALPHA A 2nd ALPHA T)	EXIT go to I/O DISP (F3) 2nd STRNG " (F1) ALPHA ALPHA E R R B = " (F1) ALPHA , ALPHA E ENTER	:DISP "ERRB=",E
	EXIT go to I/O DISP (F3) 2nd STRNG " (F1) ALPHA ALPHA E R R A =" (F1) ALPHA , ALPHA E 2nd √ (2nd CATLG-VARS go to ALL (F2) go to Σx2 ENTER ÷ 2nd CATLG-VARS go to ALL (F2) go to n ENTER) EXIT	:DISP "ERRA=",E√(ΣX2/n)
	EXIT (or 2nd QUIT)	:

CALCULATOR: TI-80 (Also for TI-82, 83, 86 with minor changes)

PROCEDURE: Regression Calculations with errors.

General Procedure:

1. Enter data. Columns 1 and 2 of sample data are shown above for linear regression (unweighted, without errors).
2. Choose type of regression calculation (linear, log, exponential, power law) and run it.
3. Run error program "LRE" (which you had entered) which gives ERRA and ERRB.
4. Use these values to calculate final errors.

Linear Regression (unweighted, with errors)

Comments	You enter	Display
Enter data and run linear regression (unweighted, without errors) as shown above.		See bottom line of linear regression program.
Run error program.	PRGM Go to LRE.	EXEC EDIT NEW
		1: LRE
	ENTER	PRGM_LRE
	ENTER	ERRB=
Error for the slope is given.		.316
		ERRA=
Error for the intercept is given.		.866
		DONE

CALCULATOR: TI-80 (Also for TI-82, 83, 86 with minor changes)

PROCEDURE: Exponential regression. This fits data to an exponentially varying curve, such as the voltage in an RC- or LR-circuit. A weight factor proportional to the square of the voltage corrects for the large relative errors for small signals.

Comments	You enter	Display
1. Data entry. Enter x (time) in L1, y (signal) in L2, and the weight W in L3. Weights should be proportional to y^2 and should be scaled and rounded off to integers.	Sample data: x(L1) y(L2) W(L3) 1 1 1 2 2 4 3 4 16 4 7 49	(Same as table to the left)
2. Do exponential regression. For final answers see below.	Press [STAT] ► Go to ExpReg. [2nd] [L1] [,] [2nd] [L2] [,] [2nd] [L3] (or [ALPHA] L 1 [,] [ALPHA] L 2 [,] [ALPHA] L 3) [ENTER]	EXPREG Y=a×b^X a=.614 b=1.842 (r^2=.995) r=.998
3. Run error program. For final answers see below.	[PRGM] Go to LRE. [ENTER] [ENTER]	ERRB= .089 ERRA= .329 DONE

Final answers for exponential regression for $y=Ae^{Bx}$ from a, b, ERRA, and ERRB, with examples:

Values: A=a=0.614 B=ln(b)=0.611

Errors: σ_A=(AB/b)ERRA= 0.067 σ_B=(B/b)ERRB=0.03

Time constant=τ=1/B=1.64. Error= $\tau\,\sigma_B$/B=0.08

CALCULATOR: TI-80 (Also for TI-82, 83, 86 with minor changes)

PROCEDURE: Exponential regression (counts). This fits data to an exponentially varying count rate, such as the decay of a radioactive isotope or the growth of a cell culture. A weight factor proportional to the count rate corrects for the large relative errors for small signals.

Comments	You enter	Display
1. Data entry. Enter x (time) in L1, y (signal) in L2, and the weight W in L3. Weights should be proportional to y and should be scaled and rounded off to integers.	Sample data: x(L1) y(L2) W(L3) 1 1 1 2 2 2 3 4 4 4 7 7	(Same as table to the left)
2. Do exponential regression. For final answers see below	Press STAT ► Go to ExpReg 2nd L1 , 2nd L2 , 2nd L3 (or ALPHA L 1 , ALPHA L 2 , ALPHA L 3) ENTER	EXPREG Y=a×b^X a=.566 b=1.885 (r^2=0.996) r=.998
3. Run error program. For final answers see below	PRGM Go to LRE ENTER ENTER	ERRB= .082 ERRA= .274 DONE

Final answers for exponential regression for $y=Ae^{Bx}$ from a, b, ERRA, and ERRB, with examples:
Values: A=a=0.566 B=ln(b)=0.634

Errors: σ_A=(AB/b) ERRA= 0.052 σ_B=(B/b)ERRB=0.028

Cell division time=$T_{1/2}$= ln(2)/B=1.09. Error=$T_{1/2}\,\sigma_B/B$=0.05. The same expressions hold for radioactive half-life, except that (-B) is used instead of B.

CALCULATOR: TI-80 (Also for TI-82, 83, 86 with minor changes)

PROCEDURE: Logarithmic regression. This fits data to a logarithmic curve, such as the potential between two cylindrical conductors. There is no weight factor.

Comments	You enter	Display
1. Data entry. Enter x (time) in L1 and y (signal) in L2.	Sample data: x(L1) y(L2) 1 1 2 2 3 4 4 7	(Same as table to the left)
2. Do logarithmic regression.	Press [STAT] ► Go to LnReg. [2nd] [L1] [,] [2nd] [L2] (or [ALPHA] L 1 [,] [ALPHA] L 2) [ENTER]	LnREG Y=a+bLnX a=.304 b=4.023 (r^2=.836) r=.914
3. Run error program.	[PRGM] Go to LRE. [ENTER] [ENTER]	ERRB= 1.262 ERRA= 1.199 DONE

CALCULATOR: TI-80 (Also for TI-82, 83, 86 with minor changes)

PROCEDURE: Power law regression. This fits data to an power law relation, such as the mass of a ball as a function of its radius. A weight factor proportional to the square of y corrects for the large relative errors for small values.

Comments	You enter	Display
1. Data entry. Enter x in L1, y in L2, and the weight W in L3. Weights should be proportional to y^2 and should be scaled and rounded off to integers between 1 and 99.	Sample data: x(L1) y(L2) W(L3) 1 1 1 2 2 4 3 4 16 4 7 49	(Same as table to the left)
2. Do power law regression. For final answers see below.	Press STAT ► Go to PwrReg. (Or 2nd STAT go to CALC (F1) MORE go to PwrR (F1) 2nd L1 , 2nd L2 , 2nd L3 (or ALPHA L 1 , ALPHA L 2 , ALPHA L 3) ENTER	EXPREG Y=aX^b a=.688 b=1.659 (r^2=.978) r=.989 ERRB= .175 ERRA= .225
3. Run error program. For final answers see below.	PRGM Go to LRE. ENTER ENTER	DONE

Final answers for power law regression for $y=Ax^B$ from a, b, ERRA, and ERRB, with examples.

Values: A=a=0.688 B=b=1.659

Errors: σ_A=A×ERRA= 0.156 σ_B=ERRB=0.175

Trigonometric Operations on the TI-80 Calculator.

This calculator is in the decimal degrees or radian mode when it is turned on.

1. To check or choose the mode, press **MODE** and go to RADIAN or DEGREES and **ENTER** **2nd** **QUIT**.

2. How to make conversions from one set of units to another:

   2a. Decimal degrees D.d to degrees, minutes, and seconds D° M' S".

   0.d ×60 **ENTER** 0.m ×60 **ENTER** (See Example below, steps 2,3.)

   2b. Degrees, minutes, and seconds D° M' S" to decimal degrees D.d.

   D+M÷60+S÷3600 (See Example below, step 4.)

   2c. Decimal degrees D.d to radians R.

   Press **MODE** and go to **RADIAN** **ENTER** **2nd** **QUIT** D.d

   2nd **ANGLE** **ENTER** **ENTER** (See Example below, step 5.)

   2d. Radians R to decimal degrees D.d.

   Press **MODE** and go to **DEGREE** **ENTER** **2nd** **QUIT** R **2nd**

   ANGLE go to **r** **ENTER** **ENTER** (See Example below, step 6.)

3. How to find the sine, cosine or tangent of an angle. (See Example below at bottom)

   3a. Given in degrees, minutes, and seconds D° M' S". Follow directions in 1 to get in degree mode, then follow 2b above to get angle in D.d. Press (sin cos or tan) D.d.

   3b. Given in decimal degrees D.d. Follow directions in 1 to get in degree mode, then press (sin cos or tan) D.d.

   3c. Given in radians R. Follow directions in 1 to get in radian mode, then press (sin cos or tan) R.

4. -Given the sine, cosine, or tangent of an angle θ, to find θ (to find $\sin^{-1}(x)$, $\cos^{-1}(x)$, or $\tan^{-1}(x)$). (See Example below at bottom.)

   4a. In radians R. Follow directions in 1 to get in radian mode, then press **2nd** (sin-1 cos-1 or tan-1) x.

   4b. In decimal degrees D.d. Follow directions in 1 to get in degree mode, then press **2nd** (sin-1 cos-1 or tan-1) x.

   4c. In degrees, minutes and seconds D° M' S". Follow directions in 1 to get in degree mode, then press **2nd** (sin-1 cos-1 or tan-1) x. Then follow directions in 2a to convert to D°M'S".

Example for the TI-80

Convert 1 7/8° to degrees, minutes, and seconds and back again. Then convert it to radians and back again.

Solution

Step	Press	Reading	Comments
1	1+7÷8 ENTER	1.875	1.875°, the angle D.d in dec. degr.
2	-1 ENTER ×60 ENTER	52.5	M.m', the number of minutes
3	-52 ENTER ×60 ENTER	30	S, the number of seconds
			D°M'S"=1°52'30"
4	1+52÷60+30÷3600 ENTER	1.875	1.875°, the angle D.d in dec. degr.
5	MODE go to RADIAN ENTER		
	2nd QUIT 1.875 2nd ANGLE		
	ENTER ENTER	0.0327249	The angle in radians
6	MODE go to DEGREE		
	ENTER 2nd QUIT 0.0327249		
	2nd ANGLE go to r		
	ENTER ENTER	1.875	1.875°, the angle D.d in dec. degr.

Example for the TI-80 . Trigonometric functions.

For θ = 1.875° = 1°52'30"=0.0327249 radian, sin(θ)=0.0327191.

For x=0.0327191, $\sin^{-1}(x)$=1.875° = 1°52'30"=0.0327249 radian

Trigonometric Operations on the TI-82, TI-83 Calculators.

This calculator is in the decimal degrees or radian mode when it is turned on.

1. To check or choose the mode, press [MODE] and go to RADIAN or DEGREES and [ENTER] [2nd] [QUIT].

2. How to make conversions from one set of units to another:

2a. Decimal degrees D.d to degrees, minutes, and seconds D° M' S".

D.d [2nd] [ANGLE] go to ►DMS [ENTER] (See Example below, step 2.)

2b. Degrees, minutes, and seconds D° M' S" to decimal degrees D.d.

TI-82: D [2nd] [ANGLE] **GO TO '** [ENTER] M [2nd] [ANGLE] **GO TO '**

[ENTER] S **GO TO '** [ENTER] [ENTER] (See Example below, step 3.)

TI-83: D [2nd] [ANGLE] [ENTER] M [2nd] [ANGLE] **GO TO '** [ENTER] S [ALPHA] "

[ENTER] (See Example below, step 3.)

2c. Decimal degrees D.d to radians R.

Press [MODE] and go to **RADIAN** [ENTER] [2nd] [QUIT] D.d

[2nd] [ANGLE] [ENTER] [ENTER] (See Example below, step 4.)

2d. Radians R to decimal degrees D.d.

Press [MODE] and go to **DEGREE** [ENTER] [2nd] [QUIT] R [2nd]

[ANGLE] go to **r** [ENTER] [ENTER] (See Example below, step 5.)

3. How to find the sine, cosine, or tangent of an angle. (See Example below at bottom.)

3a. Given in degrees, minutes, and seconds D° M' S". Follow directions in 1 to get in degree mode, then follow 2b above to get angle in D.d. Press ([sin] [cos] or [tan]) D.d.

3b. Given in decimal degrees D.d. Follow directions in 1 to get in degree mode, then press ([sin] [cos] or [tan]) D.d.

3c. Given in radians R. Follow directions in 1 to get in radian mode, then press ([sin] [cos] or [tan]) R.

4. Given the sine, cosine, or tangent of an angle θ, to find θ (to find $\sin^{-1}(x)$, $\cos^{-1}(x)$, or $\tan^{-1}(x)$). (See Example below at bottom.)

4a. In radians R. Follow directions in 1 to get in radian mode, then press [2nd] ([sin-1] [cos-1] or [tan-1]) x.

4b. In decimal degrees D.d. Follow directions in 1 to get in degree mode, then press [2nd] ([sin-1] [cos-1] or [tan-1]) x.

4c. In degrees, minutes, and seconds D° M' S". Follow directions in 1 to get in degree mode, then press [2nd] ([sin-1] [cos-1] or [tan-1]) x. Then follow directions in 2a to convert to D°M'S".

Example for the TI-82, TI-83

Convert 1 7/8° to degrees, minutes, and seconds and back again. Then convert it to radians and back again.

Solution

Step	Press	Reading	Comments
1	1+7÷8 [ENTER]	1.875	1.875°, the angle D.d in dec. degr.
2	1.875 [2nd] [ANGLE] go to		
	►DMS [ENTER]	1°52'30"	The angle in D°M'S".
3	TI-82: 1 [2nd] [ANGLE] go to '		
	[ENTER] 52 [2nd] [ANGLE] go to '		
	[ENTER] 30 go to ' [ENTER]		
	[ENTER]	1.875°	The angle D.d in dec. degr.
	TI-83: 1 [2nd] [ANGLE] [ENTER]		
	52 [2nd] [ANGLE] GO TO '		
	[ENTER] 30 [ALPHA] " [ENTER]	1.875°	The angle D.d in dec. degr.
4	[MODE] go to **RADIAN** [ENTER]		
	[2nd] [QUIT] 1.875 [2nd] [ANGLE]		
	[ENTER] [ENTER]	0.0327249	The angle in radians
5	[MODE] go to **DEGREE**		
	[ENTER] [2nd] [QUIT] 0.0327249		
	[2nd] [ANGLE] go to r		
	[ENTER] [ENTER]	1.875	1.875°, the angle D.d in dec. degr.

Example for the TI- 82, TI-83. Trigonometric functions.

For θ = 1.875° = 1°52'30"=0.0327249 radian, sin(θ)=0.0327191.

For x=0.0327191, $\sin^{-1}(x)$=1.875° = 1°52'30"=0.0327249 radian

Trigonometric Operations on the TI-86 Calculator.

This calculator is in the decimal degrees or radian mode when it is turned on.

1. To check or choose the mode, press **2nd** **MODE** and go to RADIAN or DEGREES and **ENTER** **2nd** **QUIT**.

2. How to make conversions from one set of units to another:

   2a. Decimal degrees D.d to degrees, minutes, and seconds D° M' S".

   D.d **2nd** **MATH** go to **ANGLE** (F3) go to ►DMS (F4) **ENTER** (See Example below, step 2)

   2b. Degrees, minutes, and seconds D° M' S" to decimal degrees D.d.

   D **2nd** **MATH** go to **ANGLE** (F3) go to ' (F3) M go to ' (F3) S go to ' (F3) **ENTER**. (See Example, step 3.)

   2c. Decimal degrees D.d to radians R.

   Press **2nd** **MODE** and go to **RADIAN** **ENTER** **2nd** **QUIT** D.d **2nd** **MATH** go to **ANGLE** (F3) go to ° (F1) **ENTER**. (See Example below, step 4.)

   2d. Radians R to decimal degrees D.d.

   Press **2nd** **MODE** and go to **DEGREE** **ENTER** **2nd** **QUIT** R **2nd** **MATH** go to **ANGLE** (F3) go to **r** (F2) **ENTER**. (See Example, step 5.)

3. How to find the sine, cosine or tangent of an angle. (See Example below at bottom.)

   3a. Given in degrees, minutes, and seconds D° M' S". Follow directions in 1 to get in degree mode, then follow 2b above to get angle in D.d. Press (sin, cos, or tan) D.d.

   3b. Given in decimal degrees D.d. Follow directions in 1 to get in degree mode, then press (sin, cos, or tan) D.d.

   3c. Given in radians R. Follow directions in 1 to get in radian mode, then press (sin, cos, or tan) R.

4. Given the sine, cosine, or tangent of an angle θ, to find θ [to find $\sin^{-1}(x)$, $\cos^{-1}(x)$, or $\tan^{-1}(x)$]. (See Example below at bottom.)

   4a. In radians R. Follow directions in 1 to get in radian mode, then press **2nd** (sin-1, cos-1, or tan-1) x.

   4b. In decimal degrees D.d. Follow directions in 1 to get in degree mode, then press **2nd** (sin-1, cos-1, or tan-1) x.

   4c. In degrees, minutes, and seconds D° M' S". Follow directions in 1 to get in degree mode, then press **2nd** (sin-1, cos-1, or tan-1) x. Then follow directions in 2a to convert to D°M'S".

CALCULATOR: TI-81

PROCEDURE: Entering statistical data.

Setup for display roundoff. (Permanent: do only on first use.)

1. Press [MODE] to display the MODE screen. Leave it on NORMAL.
2. Move cursor to second line, then move it to the right to the number of decimals you wish to display, for example 3, and press [ENTER], then [2nd] [QUIT].

Data entry

1. Press [2nd] [STAT], move to DATA. Optional: move to 2:ClrStat and press [ENTER], [ENTER] to erase old data and repeat previous line.
2. Move to 1: Edit and press [ENTER]. For single variable data (single entries of a single variable to calculate SD), enter data x_1, [ENTER], leave or change y1=1, enter x_2, [ENTER], etc., until all x entries are in and all y entries are equal to 1. For two variable data [calculation of standard deviation (SD) for grouped or weighted data], fill in the y-values. Press [2nd] [QUIT] and proceed to statistical calculations (see below).

CALCULATOR: TI-81

PROCEDURE: Calculation of mean, sample standard deviation (σ_{N-1}), standard error of mean (SEM).

Comments	You enter	Display
Enter single variable data (see above). Then quit or exit edit menu.	1. Press [2nd], [STAT], go to 1-Var, press [ENTER],	EDIT CALC 1: 1-VAR
Sample values are 1, 2, 3.	[ENTER],	1-VAR
Mean= $\bar{x}$=2.000		$\bar{x}$=2.000
		ΣX=6.000
		ΣX^2=14.000
Sample SD=Sx=σ_{N-1}=1.000		Sx=1.000
Population SD=σx=σ_N=0.816		σx=.816
		n=3.0000
SEM=(σ_{N-1})/√(N) = 0.577	3. 1 [÷] [2nd] [√] 3 [ENTER]	1/√3 (or 1/√(3) .577

Example for the TI-86

Convert 1 7/8° to degrees, minutes and seconds and back again. Then convert it to radians and back again.

Solution

Step	Press	Reading	Comments
1	1+7÷8 [ENTER]	1.875	1.875°, the angle D.d in dec. degr.
2	1.875 [2nd] [MATH] go to [ANGLE] (F3) go to ►DMS (F4) [ENTER]	1°52'30"	The angle in D°M'S".
3	1 [2nd] [MATH] go to [ANGLE] (F3) go to ' (F3) 52 go to ' (F3) 30 go to ' (F3) [ENTER]	1.875°	The angle D.d in dec. degr.
4	[2nd] [MODE] go to RADIAN [ENTER] [2nd] [QUIT] 1.875 [2nd] [MATH] go to [ANGLE] (F3) go to ° (F1) [ENTER]	0.0327249	The angle in radians
5	[2nd] [MODE] go to DEGREE [ENTER] [2nd] [QUIT] 0.0327249 [2nd] [MATH] go to [ANGLE] (F3) go to r (F2) [ENTER]	1.875	1.875°, the angle D.d in dec. degr.

Example for the TI-86. Trigonometric functions.

For θ = 1.875° = 1°52'30"=0.0327249 radian, sin(θ)=0.0327191.

For x=0.0327191, $\sin^{-1}$(x)=1.875° = 1°52'30"=0.0327249 radian

9/11/97

CALCULATOR: TI-81

PROCEDURE: Calculation of mean, sample standard deviation (σ_{N-1}), standard error of mean (SEM) for grouped data.

Comments	You enter	Display
Enter two-variable data, then quit or exit edit menu.	1. Press 2nd , STAT ,	CALC
Sample values are shown as entered.	ENTER, ENTER	1: 1-VAR
		1-VAR
x1=1, y1=4; x2=2, y2=18;		$\bar{x}$=3.000
x3=3, y3=24; x4=4, y4=18;		ΣX=204.000
x5=5, y5=4. Mean=$\bar{x}$=3,		ΣX^2=680.000
Sample SD=Sx=σ_{N-1}=1.007		Sx=1.007
Population SD=σx=σ_N=1.000		σx=1.000
		n=68.000
SEM=(σ_{N-1})/√(N) = 0.122	3. 1.007 ÷ 2nd √ 68	1.007/√68
	ENTER	.122

CALCULATOR: TI-81

PROCEDURE: Linear regression (unweighted, without errors), fits y=a+bx to a data set.

Comments	You enter	Display
Enter with this sample data:	1. 2nd STAT go to	EDIT CALC
(x) y)	2: LinReg	LinReg
1 1	3. Press ENTER , ENTER	
2 2		a=-1.500
3 4		b=2.000
4 7		r=.976

CALCULATOR: TI-81

PROCEDURE: Linear Regression. Program to find errors in a and b. This program, with small additions, can be used for all four types of regression (linear, log, exponential and power law).

Comments	You enter	Display
Before you can find the errors you must type in the data, this program, and do the regression calculation.	1. PRGM, go to EDIT, go to first unused Prgm, ENTER	PROGRAM NAME=
Name the prgm "LRE"	2. L,R,E, ENTER	PROGRAM:NEW
Enter the program.	3. VARS, go to LR, go to b,	PROGRAM:LRE
WARNING: Use 2nd QUIT to get out of editing. The CLEAR button will erase a whole line of the program.	ENTER, 2nd, √, (, (, VARS, go to LR, go to r, ENTER, x^{-1}, x^2, -, 1,), ÷, (, VARS, ENTER, -, 2,),), STO▶, E, ENTER	:b√((r$^{-1 2}$-1)/(n-2))->E
	ALPHA, E, 2nd, √, (, VARS go to Σ, go to Σx^2 ENTER, ÷, VARS, ENTER,), STO▶, F, ENTER	:E√(ΣX^2/n)->F
	PRGM, go to I/O, ENTER, 2nd, A-LOCK, ", E, R, R, ␣ (above 0), A, ",", B, 2nd, TEST, ENTER, ALPHA, ", ENTER	:DISP "ERR A,B="
	PRGM go to I/O, ENTER, ALPHA, F, ENTER	:DISP F
	PRGM, go to I/O, ENTER, ALPHA, E, 2nd, QUIT	:Disp E

CALCULATOR: TI-81

PROCEDURE: Regression Calculations with errors.

General Procedure:

1. Enter data. Columns 1 and 2 of sample data are shown above for linear regression (unweighted, without errors).
2. Choose type of regression calculation (linear, log, exponential, power law) and run it.
3. Run error program "LRE" (which you had entered) which gives ERRA and ERRB.
4. Use these values to calculate final errors.

Linear Regression (unweighted, with errors)

Comments	You enter	Display
Enter data and run linear regression (unweighted, without errors) as shown above.		See bottom line of linear regression program.
Run error program, which may have a different number than 1,	[PRGM] If necessary, go to LRE.	**EXEC** EDIT NEW **1:** Prgm1 LRE Prgm1 (or other number for Pgrm)
Errors for the slope and intercepts are given.	[ENTER] [ENTER]	ERR A,B= .866 .316

CALCULATOR: TI-81
PROCEDURE: Exponential regression. This fits data to an exponentially varying curve, such as the voltage in an RC or LR circuit, nuclear decay, or cellular growth. Since no weight factor is possible, the results for A,B will differ slightly, the errors more so, from weighted fits.

Comments	You enter	Display
1. Data entry. Enter x (time), y (signal). Sample data are as above.	Press [2nd], [STAT]. Go to	
2. Do exponential regression. (Prgm can have a number different from 1.)	ExpReg, [ENTER]	ExpReg a=.535 b=1.921 r=.999
	[PRGM] Go to LRE. [ENTER]	Prgm 1
3. Run error program. For final answers see below.	[ENTER]	ERR A,B= .186 .068

Final answers for exponential regression for $y=Ae^{Bx}$ from a, b, ERRA, and ERRB, with examples:
Values: A=a=0.535 B=ln(b)=0.653

Errors: σ_A=(AB/b)ERRA= 0.034 σ_B=(B/b)ERRB=0.023

Time constant=τ=1/B=1.53. Error= $\tau\,\sigma_B$/B=0.05. Cell division time=$T_{1/2}$= ln(2)/B=1.06. Error=$T_{1/2}\,\sigma_B$/B=0.04. The same expressions hold for radioactive half-life, except that (-B) is used instead of B.

CALCULATOR: TI-81

PROCEDURE: Logarithmic regression. This fits data to a logarithmic curve, such as the potential between two cylindrical conductors, of the form y=a+b*ln(x). There is no weight factor.

Comments		Display
1. Sample data are as above.		
2. Do logarithmic regression.	Press [2nd], [STAT], go to LnReg, [ENTER], [ENTER]	LnREG a=.304 b=4.023 r=.914
	[PRGM] Go to LRE. [ENTER]	ERR A,B=
3. Run error program. (Number may be different from 1.)	[ENTER]	1.199 1.262

CALCULATOR: TI-81
PROCEDURE: Power law regression. This fits data to an power law relation, such as the mass of a ball as a function of its radius. Y=aX^b Since no weight factor is possible, the results for a,b will differ slightly, the errors more so, from weighted fits.

Comments	You enter	Display
1. Sample data are as above.		
2. Do power law regression. For final answers see below. (Prgm number may differ from 1.)	Press **2nd**, **STAT**, go to PwrReg, **ENTER**, **ENTER**	PwrReg a=.909 b=1.386 r=.987
	PRGM Go to LRE. **ENTER**	Prgm1 ERR A,B=
3. Run error program. For final answers see below.	**ENTER**	.151 .159

Final answers for power law regression for $y=Ax^B$ from a, b, ERRA, and ERRB, with examples.
Values: A=a=0.909 B=b=1.386
Errors: $\sigma_A = A \times ERRA = 0.137$ $\sigma_B = ERRB = 0.159$

Trigonometric Operations on the TI-81 Calculator.

This calculator is in the decimal degrees or radian mode when it is turned on.

1. To check or choose the mode, press **MODE** and go to RADIAN or DEGREES and **ENTER** **2nd** **QUIT**.

2. How to make conversions from one set of units to another:

    2a. Decimal degrees D.d to degrees, minutes and seconds D° M' S".

    0.d ×60 **ENTER** 0.m ×60 **ENTER** (See Example below, steps 2,3.)

    2b. Degrees, minutes, and seconds D° M' S" to decimal degrees D.d.

    D+M÷60+S÷3600 (See Example below, step 4.)

    2c. Decimal degrees D.d to radians R.

    Press **MODE** and go to **RADIAN** **ENTER** **2nd** **QUIT** D.d

    MATH go to °, **ENTER** **ENTER** (See Example below, step 5.)

    2d. Radians R to decimal degrees D.d.

    Press **MODE** and go to **DEGREE** **ENTER** **2nd** **QUIT** R

    MATH go to **r** **ENTER** **ENTER** (See Example below, step 6.)

3. How to find the sine, cosine, or tangent of an angle. (See Example below.)

    3a. Given in degrees, minutes and seconds D° M' S". Follow directions in 1 to get in degree mode, then follow 2b above to get angle in D.d. Press (sin, cos, or tan) D.d.

    3b. Given in decimal degrees D.d. Follow directions in 1 to get in degree mode, then press (sin cos or tan) D.d.

    3c. Given in radians R. Follow directions in 1 to get in radian mode, then press (sin, cos, or tan) R.

4. Given the sine, cosine or tangent of an angle θ, to find θ [to find $\sin^{-1}(x)$, $\cos^{-1}(x)$, or $\tan^{-1}(x)$]. (See Example below at bottom.)

    4a. In radians R. Follow directions in 1 to get in radian mode, then press **2nd** (sin-1, cos-1, or tan-1) x.

    4b. In decimal degrees D.d. Follow directions in 1 to get in degree mode, then press **2nd** (sin-1, cos-1, or tan-1) x.

    4c. In degrees, minutes, and seconds D° M' S". Follow directions in 1 to get in degree mode, then press **2nd** (sin-1, cos-1, or tan-1) x. Then follow directions in 2a to convert to D°M'S".

Example for the TI-81

Convert 1 7/8° to degrees, minutes, and seconds and back again. Then convert it to radians and back again.

Solution. Start in degree mode, float, 8.

Step	Press	Reading	Comments
1	1+7÷8 ENTER	1.87500000	1.875°, the angle D.d in dec. degr.
2	-1 ENTER ×60 ENTER	52.5	M.m', the number of minutes
3	-52 ENTER ×60 ENTER	30.00000000	S, the number of seconds.
			D°M'S"=1°52'30"
4	1+52÷60+30÷3600 ENTER	1.87500000	1.875°, the angle D.d in dec. degr.
5	MODE go to RADIAN ENTER		
	2nd QUIT 1.875 MATH, go to °,		
	ENTER ENTER	0.03272492	The angle in radians
6	MODE go to DEG		
	ENTER 2nd QUIT	0.03272492	
	MATH go to r		
	ENTER ENTER	1.87500000	1.875°, the angle D.d in dec. degr.

Example for the TI-81. Trigonometric functions.

For θ = 1.875° = 1°52'30"=0.03272492 radian, go to DEG mode, press SIN,1.875, ENTER. Result is sin(θ)=0.03271908.

For x=0.03271908, press 2nd SIN^{-1}, 0.03271908, ENTER. The result is=$\sin^{-1}(x)$ =1.87499984° = 1°52'29.99941766"=0.03272492 radian.

CALCULATOR: TI-85

PROCEDURE: Entering statistical data.

Set-up for display roundoff. (Permanent: do only on first use.)

1. Press [2nd] [MODE] to display the MODE screen. Leave it on NORMAL.
2. Move cursor to second line, then move it to the right to the number of decimals you wish to display, for example 3, and press [ENTER], then [2nd] [QUIT].

Data entry

1. Press [STAT], F2 (EDIT), [ENTER], [ENTER]. Optional: press F5 (CLRxy) to erase old data.
2. For single variable data (single entries of a single variable to calculate SD), enter data x_1, [ENTER], leave or change y1=1, enter x_2, [ENTER], etc., until all x entries are in and all y entries are equal to 1. For two-variable data [calculation of standard deviation (SD) for grouped or weighted data, or regression], fill in the y-values.
3. Press [2nd], F1 (CALC on upper row of display) or F1 (if CALC is on the lower row), [ENTER], [ENTER] and proceed to statistical calculations (see below).

CALCULATOR: TI-85

PROCEDURE: Calculation of mean, sample standard deviation (SD) σ_{N-1}, standard error of mean (SEM).

Comments	You enter	Display
Enter single-variable data and enter CALC menu (see above).	1. Press F1 (1-VAR)	$\bar{x}$ =2.000
		Σx=6.000
Sample values are 1, 2, 3.		Σx^2=14.000
Mean= $\bar{x}$ =2.000		Sx=1.000
Sample SD=Sx=σ_{N-1}=1.000		σx=.816
Population SD=σx=σ_N=0.816	2. [2nd] QUIT,1 [÷] [2nd] [√] 3[ENTER]	n=3.0000
SEM=(σ_{N-1})/√(N) = 0.577		1/√3 (or 1/√(3) .577

CALCULATOR: TI-85

PROCEDURE: Calculation of mean, sample standard deviation (σ_{N-1}), standard error of mean (SEM) for grouped data.

Comments	You enter	Display

Enter two variable data, and enter CALC menu. (See Data Entry 3.).
Sample values are shown as entered
x1=1, y1=4; x2=2, y2=18;
x3=3, y3=24; x4=4, y4=18;
x5=5, y5=4. Mean=$\bar{x}$=3,
	1. Press F1 (1-VAR)	$\bar{x}$=3.000
		ΣX=204.000
		ΣX^2=680.000
		Sx=1.007
		σx=1.000
		n=68.000

Sample SD=Sx=σ_{N-1}=1.007 Population SD=σx=σ_N=1.000	2. **2nd** , QUIT, 1.007 **÷** **2nd** **√** 68 **ENTER**	1.007/ √68 .122
SEM=(σ_{N-1})/√(N) = 0.122		

CALCULATOR: TI-85

PROCEDURE: Linear Regression (unweighted, without errors), fits y=a+bx to a data set.

Comments	You enter	Display
Enter with this sample data: (x) y) 1 1 2 2 3 4 4 7 and enter CALC menu. (See Data Entry 3.)	Press F2 (LINR)	LinR a=-1.500 b=2.000 corr=.976 n=4.000

CALCULATOR: TI-85

PROCEDURE: Linear Regression. Program to find errors in a and b. This program, with small additions, can be used for all four types of regression (linear, log, exponential, and power law).

Comments	You enter	Display
Before you can find the errors you must type in the data, this program, and do the regression calculation.	1. PRGM, F2 (EDIT),	PROGRAM
	2. E, R, R, S, ENTER	NAME= ERRS
Name the prgm "ERRS".	3. 2nd, alpha, B, √, (, (,	
Enter the program.	2nd, VARS, go to corr,	:b√((corr-1 2-1)/(n-2))
WARNING: Use 2nd QUIT to	ENTER, 2nd, x-1, x2,	->E
	-,1,), ÷, (, 2nd alpha,	:DISP "ERRB=",E
get out of editing. The CLEAR button will erase a whole line of the program.	N, -, 2,),), STO▶, E,	:DISP "ERRA=",E√(ΣX2/ n)
	ENTER, F3 (I/O), F3 (Disp),	
	2nd, STRNG, F1 ("), ALPHA,	
	ALPHA, E, R, R, B, =, F1 ("),	
	ALPHA, ,, ALPHA, E,	
	ENTER, PRGM, 2nd, F3	
	(Disp), 2nd, STRNG, F1 ("),	
	ALPHA, ALPHA, E, R, R, A,	
	=, F1 ("), ALPHA, ,,	
	ALPHA, E, 2nd, √, (	
	2nd, VARS, F1 (All),	
	go to Σx2, ENTER, ÷, 2nd,	
	alpha, n,), 2nd, QUIT	

CALCULATOR: TI-85

PROCEDURE: Regression Calculations with errors.

General Procedure:

1. Enter data. Columns 1 and 2 of sample data are shown above for linear regression (unweighted, without errors).
2. Choose type of regression calculation (linear, log, exponential, power law) and run it.
3. Run error program "ERRS" (which you had entered) which gives ERRB and ERRA.
4. Use these values to calculate final errors. Sample values are for linear regression.

Linear Regression (unweighted, with errors)		
Comments	You enter	Display
Enter data, enter CALC menu (see Data Entry 3) and run linear regression (unweighted, without errors) as shown above.	2nd, QUIT, PRGM, F1 (NAMES), F key under ERRS,	NAMES EDIT
Run error program.	ENTER	ERRS ERRB= .316 ERRA= .866 DONE
Errors for the slope and intercepts are given.		

CALCULATOR: TI-85
PROCEDURE: Exponential regression. This fits data to an exponentially varying curve, such as the voltage in an RC- or LR-circuit, nuclear decay, or cellular growth. Since no weight factor is possible, the results for A,B will differ slightly, the errors more so, from weighted fits.

Comments	You enter	Display
1. Data entry. Enter x (time), y (signal). Sample data are as above. Enter CALC menu. (See Data Entry 3.) 2. Do exponential regression.	Press F4 (EXPR)	ExpR a=.535 b=1.921 corr=.999 n=4.000
3. Run error program. For final answers see below.	2nd, QUIT, PRGM, F1(NAMES), F key under ERRS, ENTER, ENTER	ERRS ERRB= .068 ERRA= .186 DONE

Final answers for exponential regression for $y=Ae^{Bx}$ from a, b, ERRA, and ERRB, with examples:
Values: A=a=0.535 B=ln(b)=0.653
Errors: σ_A=(AB/b)ERRA= 0.034 σ_B=(B/b)ERRB=0.023
Time constant=τ=1/B=1.53. Error= $\tau\,\sigma_B$/B=0.05. Cell division time=$T_{1/2}$= ln(2)/B=1.06.
Error=$T_{1/2}\,\sigma_B$/B=0.04. The same expressions hold for radioactive half-life, except that (-B) is used instead of B.

CALCULATOR: TI-85
PROCEDURE: Logarithmic regression. This fits data to a logarithmic curve, such as the potential between two cylindrical conductors, of the form y=a+b*ln(x). There is no weight factor.

Comments		Display
1. Sample data are as above. Enter CALC menu. (See Data Entry 3.) 2. Do logarithmic regression.	Press F3 (LNR)	LnR a=.304 b=4.023 r=.914 n=4.000
3. Run error program.	2nd, QUIT, PRGM, press F1 (NAMES), press F key under ERRS, ENTER	ERRS ERRB= 1.262 ERRB= 1.199

CALCULATOR: TI-85
PROCEDURE: Power law regression. This fits data to an power law relation, such as the mass of a ball as a function of its radius. Y=aX^b Since no weight factor is possible, the results for a,b will differ slightly, the errors more so, from weighted fits.

Comments	You enter	Display
1. Sample data are as above. Enter CALC menu. (See Data Entry 3.)	Press F5 (PWR).	PwrReg
2. Do power law regression.		a=.909 b=1.386 corr=.987 n=4.00
3. Run error program. For final answers see below.	**2nd**, QUIT, **PRGM**, press F1 (NAMES), press F key under ERRS, **ENTER**	ERRS ERRB= .159 ERRB= .151 DONE

Final answers for power law regression for $y=Ax^B$ from a, b, ERRA, and ERRB, with examples.
Values: A=a=0.909 B=b=1.386
Errors: σ_A=A×ERRA= 0.137 σ_B=ERRB=0.159

Trigonometric Operations on the TI-85 Calculator.

This calculator is in the decimal degrees or radian mode when it is turned on.

1. To check or choose the mode, press [2nd] [MODE] and go to RADIAN or DEGREE and [ENTER] [2nd] [QUIT].

2. How to make conversions from one set of units to another:

   2a. Decimal degrees D.d to degrees, minutes, and seconds D° M' S".

   [2nd], MATH, F3 (ANGLE), F4 (>DMS), [ENTER] (See Example below, step 2.)

   2b. Degrees, minutes, and seconds D° M' S" to decimal degrees D.d.

   To enter an angle in D° M' S", press [2nd], MATH, F3 (ANGLE), D, F3 ('), M, F3 ('), S, F3 ('). To change to D.d, press [ENTER]. If that doesn't work, get in MATH menu (press [2nd], MATH if necessary) press F1 (°), [ENTER]. (See Example below, step 3.)

   2c. Decimal degrees D.d to radians R.

   Press [2nd], [MODE] and go to **RADIAN** [ENTER] [2nd] [QUIT], D.d, [2nd], [MATH], F3 (ANGLE), F1 (°), [ENTER]. (See Example below, step 4.)

   2d. Radians R to decimal degrees D.d.

   Press [2nd], [MODE] and go to **DEGREE** [ENTER] [2nd] [QUIT], R, [2nd], [MATH], F3 (ANGLE), F2 (r), [ENTER] (See Example below, step 5.)

3. How to find the sine, cosine, or tangent of an angle. (See Example below.)

   3a. Given in degrees, minutes and seconds D° M' S". Follow directions in 1 to get in degree mode, Press ([sin], [cos], or [tan]), D° M' S" (follow 2b above to enter angle in D° M' S"), [ENTER].

   3b. Given in decimal degrees D.d. Follow directions in 1 to get in degree mode, then press ([sin], [cos], or [tan]) D.d. (Or [2nd], ANS if D.d is in display), [ENTER].

   3c. Given in radians R. Follow directions in 1 to get in radian mode, then press ([sin] [cos] or [tan]), R, [ENTER].

4. Given the sine, cosine, or tangent of an angle θ, to find θ (to find $\sin^{-1}(x)$, $\cos^{-1}(x)$, or $\tan^{-1}(x)$). (See Example below.)

   4a. In radians R. Follow directions in 1 to get in radian mode, then press [2nd] ([sin-1], [cos-1], or [tan-1]), x, [ENTER].

   4b. In decimal degrees D.d. Follow directions in 1 to get in degree mode, then press [2nd] ([sin-1], [cos-1], or [tan-1]), x, [ENTER].

   4c. In degrees, minutes, and seconds D° M' S". Follow directions in 1 to get in degree mode, then press [2nd] ([sin-1], [cos-1], or [tan-1]), x, [ENTER]. Then follow directions in 2a to convert to D°M'S".

Example for the TI-85

Convert 1 7/8° to degrees, minutes, and seconds and back again. Then convert it to radians and back again.

Solution . Start in degree mode, float, 8.

Step	Press	Reading	Comments
1	1+7÷8 ENTER	1.87500000	1.875°, the angle D.d in dec. degr.
2	2nd, MATH, F3 (ANGLE),		
	F4 (>DMS), ENTER	1° 52' 30"	The angle in D° M' S"
3	F1 (°), ENTER	1.87500000	1.875°, the angle D.d in dec. degr.
4	2nd, MODE go to RADIAN,		
	ENTER, 2nd QUIT, 2nd,		
	MATH, F3 (ANGLE), F1 (°),		
	ENTER	0.03272492	The angle in radians.
5	2nd, MODE, go to DEG		
	ENTER, 2nd, QUIT,		
	2nd, MATH, F3 (ANGLE), F2 (r),		
	ENTER	1.87500000	1.875°, the angle D.d in dec. degr.

Example for the TI-85 . Trigonometric functions. For θ = 1.875° = 1°52'30"=0.03272492 radian, go to DEG mode, press SIN, 1.875, ENTER. Result is sin(θ)=0.03271908.

To reverse the operation, press 2nd SIN-1, 2nd, ANS, ENTER. The result is=$\sin^{-1}(x)$ =1.87500000° = 1°52'30"=0.03272492 radian.

Appendix C:

DETAILS OF REGRESSION COMPUTATIONS

TABLE C-1 Estimation of Weight of Bottles by Linear Regression

Number of Units	Total Weight	x Deviation	Square of x Deviation	y Deviation	Product of Deviations	Calculated y_c	Square of Deviation
x	y (lb)	$x - \bar{x}$	$(x - \bar{x})^2$	$(y - \bar{y})$ (lb)	$(x - \bar{x})(y - \bar{y})$ (lb)	$a + bx$ (lb)	$(y - a - bx)^2$ (lb)^2
0	207.5	−2.5	6.25	−3	7.5	207.4	0.01
	207	−2.5	6.25	−3.5	8.75	207.4	0.16
1	209	−1.5	2.25	−1.5	2.25	208.64	0.13
	208	−1.5	2.25	−2.5	3.75	208.64	0.41
2	210.5	−0.5	0.25	0.0	0.0	209.88	0.38
	210.5	−0.5	0.25	0.0	0.0	209.88	0.38
3	211	+0.5	0.25	+0.5	0.25	211.12	0.01
	211	+0.5	0.25	+0.5	0.25	211.12	0.01
4	212	+1.5	2.25	+1.5	2.25	212.36	0.13
	212.5	+1.5	2.25	+2.0	3.0	212.36	0.02
5	213	+2.5	6.25	+2.5	6.25	213.6	0.36
	214	+2.5	6.25	+3.5	8.75	213.6	0.16

$\Sigma x = 30$ $\Sigma y = 2526$ lb $\Sigma(x - \bar{x})^2 = 35$ $\Sigma(x - \bar{x}(y - \bar{y}) = 43$ lb $\Sigma(y - a - bx)^2 = 2.16$ lb^2
$N = 12$ $\bar{x} = \Sigma\, x/N = 2.5$ $\bar{y} = \Sigma\, y/N = 210.5$ lb $\sigma_x = \sqrt{\Sigma\,(x - \bar{x})^2/N} = 1.71$
$\sigma = \sqrt{\Sigma\,(y - a - bx)^2/(N - 2)}; = 0.46$ lb
$b = \Sigma\,(x - \bar{x})(y - \bar{y})/(N\sigma_x^2) = 1.23$ lb $a = \bar{y} - b\bar{x} = 207.4$ lb

$\sigma_b = \dfrac{\sigma}{\sigma_x\sqrt{N}} = 0.078$, rounded to 0.08 $\sigma_a = \sigma_b\sqrt{\sigma_x^2 + \bar{x}^2} = 0.24$, rounded to 0.2
$a = 2.07.4(2)$ lb $b = 1.23(8)$ lb

TABLE C-2 Nettie's Oscilloscope Data for Measuring Time Constant RC

x (ms)	V (cm)	w (V^2)	y (ln V)	wx (msec)	wx^2 (msec^2)	wy	wxy (msec)	$d = y - a - bx$	wd^2
0	10.0	100.00	2.303	0.00	0.00	230.26	0.00	0.0017	0.0003
1	9.0	81.00	2.197	81.00	81.00	177.98	177.98	0.0056	0.0025
2	8.2	67.24	2.104	134.48	268.96	141.48	282.96	−0.0027	0.0005
3	7.4	54.76	2.001	164.28	492.84	109.60	328.80	−0.0015	0.0001
4	6.8	46.24	1.917	184.96	739.84	88.64	354.55	−0.0184	0.0156
5	6.0	36.00	1.792	180.00	900.00	64.50	322.52	0.0053	0.0010
6	5.4	29.16	1.686	174.96	1049.76	49.18	295.05	0.0093	0.0025
7	5.0	25.00	1.609	175.00	1225.00	40.24	281.65	−0.0152	0.0058
8	4.4	19.36	1.482	154.88	1239.04	28.68	229.47	0.0112	0.0024
9	4.0	16.00	1.386	144.00	1296.00	22.18	199.63	0.0051	0.0004
10	3.6	12.96	1.281	129.60	1296.00	16.60	166.01	0.0090	0.0011
	Sum:	487.72		1523.16	8588.44	969.34	2638.62		0.0323

Average: $\bar{x} = \Sigma wx/\Sigma w = 3.123$ msec $\quad \bar{y} = \Sigma wy/\Sigma w = 1.99$

$$b = \frac{(\Sigma wxy/\Sigma w) - \overline{xy}}{\sigma_x^2} = -0.101,4\ \text{msec}^{-1}, \qquad \sigma_x = \sqrt{\Sigma wx^2/\Sigma w - \bar{x}^2} = 2.80\ \text{msec}$$

$$a = \bar{y} - b\bar{x} = 2.304, \qquad V_0 = e^a = 10.0\ \text{cm}, \quad RC = -1/b = 9.86\ \text{msec}$$

$$\sigma_b = \frac{\sigma}{\sigma_x\sqrt{N-2}} = 0.000,97\ \text{msec}^{-1}, \qquad \sigma = \sqrt{\frac{\Sigma wd^2}{\Sigma w}} = 0.0081, \quad N = 11$$

$$\sigma_a = \sigma_b\sqrt{\Sigma wx^2/\Sigma w} = 0.004, \qquad \sigma_t/t = 0.01\ (1\%)$$

$$\frac{\sigma_{RC}}{RC} = \sqrt{\left(\frac{\sigma_b}{b}\right)^2 + \left(\frac{\sigma_t}{t}\right)^2} = 0.014\ (1.4\%), \qquad \sigma_{V_0} = V_0\sigma_a = 0.04\ \text{cm}$$

TABLE C-3 Zeke Measures the Mean Life of a Radioactive Isotope.

x (min)	$C = w$ (cnts = wgt)	y (ln C)	wx (min)	wx^2 (min^2)	wx	wxy (min)
0	1032	6.939	0	0	7161.0	0
10	592	6.384	5920	59,200	3779.3	37,790
20	375	5.927	7500	150,000	2222.6	44,452
30	223	5.407	6690	200,700	1205.8	37,174
40	116	4.754	4640	185,600	551.5	22,057
50	88	4.477	4400	220,000	394.0	19,700
60	40	3.689	2400	144,000	147.6	8,853
70	35	3.555	2450	171,500	124.4	8,711
Sum:	2,501		34,000	1,131,000	15,586.2	177,737

Average: $\overline{x} = \Sigma\, wx/\Sigma\, w = 13.6$ min, $\overline{y} = \Sigma\, wy/\Sigma\, w = 6.232$

$$b = \frac{(\Sigma\, wxy/\Sigma\, w) - \overline{xy}}{\sigma_x^2} = -0.051 \text{ min}^{-1}, \qquad \sigma_x = \sqrt{\frac{\Sigma\, wx^2}{\Sigma\, w} - \overline{x}^2} = 16.35 \text{ min}$$

$$a = \overline{y} - b\overline{x} = 6.923, \qquad C_0 = e^a = 1020, \qquad \tau = -1/b = 19.6$$

$$\sigma_b = 1/(\sigma_x \sqrt{(\Sigma\, w)}) = 0.0012 \text{ min}^{-1}, \qquad \sigma_a = \sigma_b \sqrt{(\Sigma\, wx^2/\Sigma\, w)} = 0.026$$

$$\sigma_\tau = \sigma_b/b^2 = 0.5 \text{ min}, \qquad \sigma_{C_0} = C_0\sigma_a = 30$$

Appendix D:

DESCRIPTION AND INSTRUCTIONS FOR EXCEL™ WORKSHEETS FOR DATA FITS CT ERROR ANALYSIS PROGRAMS

EXCEL™ WORKSHEETS FOR DATA FITS. DESCRIPTION, AND INSTRUCTIONS

These programs find several types of "best" fits to a data set. The programs include simple or grouped averages, straight line, exponential, power law, logarithmic, and polynomial fits.

For best appearance, use the Windows version on a PC running Microsoft Excel on Windows 95 or Windows 3.1, and the Mac Versions on a Macintosh computer running Excel. All worksheets except FitHaven will work on Excel version 4 as well as the later Excel 5 or current Microsoft Office versions. There may be difficulty in converting to Lotus 1-2-3 or Quattro-Pro, which may differ from Excel in some worksheet functions.

The FitHaven workspace will only work on Excel 5 or later versions. It features error checking and graphical error bars not found on the earlier worksheets and performs all of their fitting functions in a single program. If your machine does not have enough memory to run FitHaven, use the other worksheets which each perform a single type of fit.

INSTRUCTIONS FOR SINGLE-CONCEPT WORKSHEETS

IN FOLDER MacXL or DIRECTORY PcXL

Windows Versions	Mac Versions
Fit Avg.xls	FitAverage.XL4
(Find averages and standard deviations)	
FitGroup.xls	FitGroupAverage.XL4
(Find averages of grouped data)	
FitLine.xls	FitLine.XL4
(Find slope/intercept of line fitting XY data)	
FitLino.xls	FitLineThroughOrigin.XL4
(Find slope of line passing through origin)	
FitExp.xls	FitExponential.XL4
(Finds best fit to exponential curve)	
FitCount.xls	FitExpToCount.XL4
(Finds best exponential decay fitting count data)	
FitPwr.xls	FitPowerLaw.XL4
(Finds best fit power law curve)	
FitLog.xls	FitLog.XL4
(Finds best fit logarithmic curve)	
FitPoly.xls	FitPolynomial.XL4
(Finds best fit polynomial curve to XY data)	

Start your copy of Microsoft Excel.

1. Open the desired worksheet.
2. If the sheet appears too large for your screen, select the WINDOWS: ARRANGE:TILE menu option in Excel.
3. Type in the values of your Y or X/Y data values in the left-hand columns.
4. Fitted parameters appear at top center, with graph below. You may scroll to right or down to find additional results and instructions.
5. To save your the worksheet on your hard disk, use a new file name. This will leave the original program available for future use. Avoid using locked or read-only files, which may cause problems when data is cleared. Have a backup copy of the program for reloading, in case the working copy gets corrupted.

INSTRUCTIONS FOR FitHaven WORKSPACE

1. Start your Microsoft Excel program.
2. Open the FitHaven.XL5 file from the XL_MAC folder on a Macintosh, or the FitHaven.xls file from the XL_PC folder on a Windows 95 or Windows 3.1 machine. (The program will both work on either machine, but has been designed for best appearance on the designated machines.)
3. If the sheet appears too large for your screen, select the Windows:Arrange:Tile menu option in Excel.
4. Select the desired worksheet by clicking on the tabs at the bottom. Instructions are on the first sheet.
5. Type in the values of your Y or X/Y data values in the left-hand columns.
6. Click on the button for the desired fit.
7. Fitted parameters appear at top center, with graph below. You may scroll to right or down to find additional results and instructions.
8. To save your the worksheet on your hard disk, use a new filename. This will leave the original program available for future use. Avoid using locked or read-only files, which may cause problems when data is cleared. Have a backup copy of the program for reloading, in case the working copy gets corrupted.

cT Error Analysis Programs. Description and Instructions.

There are two cT error analysis programs:

1. AddErr.ctb is a tutorial on Chapter 3 (Error Analysis for More Than One Variable). A graphic demonstration of the random walk is included.
2. The ErrProp.ctb program calculates the error in a function of more than one random variable.

Instructions

1. Macintosh Version: Copy the contents of the StatMac folder to your hard disk (program cT_Executor, files commands.tab, function.tab, ErrProp.ctb, and AddErr.ctb).

   Windows 95 Version: Start Windows 95. Copy the contents of the StatPC folder (program cTwinX.exe and the lessons ErrProp.ctb and AddErr.ctb) to a subdirectory on your hard disk.
2. Run the cT program and open the file AddErr.ctb for the tutorial and demonstrations. Run the cT program and open the file ErrProp.ctb to do error analysis.
3. Navigational hints are given inside the lessons.
4. If you are taking a lesson for the first time, select the Introduction to learn how to move around inside the lesson and how to answer questions.
5. Select QUIT from the option menu to leave the lesson.

Appendix E:

ANSWERS TO PROBLEMS

Chapter 1

Answers to Problems

1-1. (a) exact (b) approximate (c) exact (d) exact (e) exact (f) approximate
1-2. 133 lb
1-3. (a) 41.3 cm (b) 30.4 cm (c) Yes (d) No
1-4. (a) 31.5 lb (b) 164 lb (c) No.
1-5. 12.1(5) cm (b) 234.67 ± 0.03 g (c) 983.5 ± 1.2 g (d) 56.23(17) g
1-6. (a) 2 mph (b) 2 mph (c) 4% (d) 40% (e) 100 mph; 2%
1-7. (a) 261.7(3) g (b) 63.29(1) g (c) 1030(10)g
1-8. (a) 50 g (b) 25 g
1-9. 176.4 ± 0.2 cm (b) 1.764 ± 0.002 m (c) 69.43 ± 0.08″
1-10. 0.63 s (b) 0.04 s (c) 0.04 s (d) 0.07 (e) 7% (f) 63.4(4) s, 63.4 ± 0.04 s
1-11. 0.0005 sec, or 2%
1-12. (a) 27 mg/ml (b) 30 ± 3 mg/ml

Chapter 2

Answers to Problems

2-1. (a) 5.2 (b) 0.44s (c) 0.44s (d) 0.41 − 0.48 s (e) 0.02 s (f) 0.012 s (g) 0.03 (h) 3%
2-2. (a) 5 (b) 2.7 (c) 5.3 (d) 2.4
2-3. (a) 20.5″ (b) 4.0″ (c) 22.4″, 20.8″, 18.2″, 18.6″, 22.4″ (d) 1.8″ (d) 1.8″
2-4. (a) 68% (b) 38% (c) 0 (d) 1.2% (e) 0.13%
2-5. (a) 2/1 against (b) even odds (c) 5 to 1 against (d) 100/1 against
2-6. (a) 0.03 s (b) 1.5% (c) 0.5% (d) 2.5% (e) 0.25%
2-7. (a) 4 (b) 100
2-10. (a) 0.58 (b) 0.65
2-11. (a) 3.14″ (b) 2.8″ (c) 3.2″ (d) 2.7″
2-12. 12.7
2-13. 5%

Chapter 3

Answers to Problems

3-1. (a) 1% (b) 3%
3-2. (a) 4.004(4) In^2 (b) 0.1%
3-3. Hole depth $= 99.0 \pm 0.1 \pm 0.099$ cm; to step top $98.0 \pm 0.1 \pm 0.098$ cm (same sign as for the hole); step $1.0 \pm 0.14 \pm 0.001$ cm. (The first error is random; the second is systematic.)
3-4. 0.000 003 cm
3-5. (a) 0.1 cm (b) 0.03 (c) 3% (d) 0.030(1) cm (e) 0.03 (f) 3%
3-6. (a) 0.5 g/cm^3 (b) 5% (c) 5%
3-7. (a) 7.74(6) g/cm^3 (b) 0.008 (c) 0.008
3-8. (a) 0.001 (b) 0.022 (c) 9.9 m/s^2 (d) 0.04 (e) error in y negligible.
3-9. 10%
3-10. (a) 60 m (b) 11 days 14 hours
3-11. $\beta = 3\alpha$
3-12. 9×10^{-6} °C^{-1}
3-13. (a) $\frac{1}{273} = 3.7 \times 10^{-3}$ °C^{-1} (b) $\beta = \beta'$ (c) $-\frac{1}{4}$%

Chapter 4

Answers to Problems

4-1. (a) 34.5 cm/s (b) 1.7 cm/sec (c) 0.55 s (d) −19 cm
4-2. (a) b $= 0.0697 \pm 0.0010$ m/kg (b) Yes. Delete the first point. (c) b $= 0.0713 \pm 0.0001$
4-2. (a) −3.5 kg (−34 N) (b) 0.139 kg/cm (13.7 N/cm) (c) 25 cm (d) 32 cm
4-3. a $=$ 3.000(1124); b $=$ 0.500(118) in all cases.
4-4. 3.92(8) cm/sec^2
4-5. 3.8(5) cm/sec^2
4-6. Women. Linear: a $= -109$, b $=$ 3.69(6), midget: 2 lb
Quadratic: d $= 10.2$, c $=$ 0.0284(4), midget: 36 lb
Men. Linear: a $= -168$, b $=$ 4.80(6), midget: −24 lb
Quadratic: d $= -3.4$, c $=$ 0.0348(3), midget: 28 lb
4-7. L $=$ 38(1) H; C $=$ 2.14(9) μF
4-9. 0.0312(3) m/kg
4-10. (a) $a = 0.03 \pm 0.01$ (sec^2/cm); $b = 0.0238 \pm 0.0004$ (sec^2/g·cm) (b) $m = 3a/\text{b}$ $= 4.1 \pm 1.6$ g $= 0.0041 \pm 0.0016$ kg (c) $k = 4\pi^2/b = 1660 \pm 30$ (g·cm/sec^2) $= 1.66 \pm 0.03$ N/m
4-11. (a) 0.95 (b) 0.7 (c) 1.2
4-12. (a) −250°C (b) 100°C (c) 30°C

Chapter 5

Answers to Problems

5-1. (a) 360° (b) 6.283

5-2. (a) 1° 56′ (b) 1.933° (c) 0.0337 (d) 0.0337

5-3. (a) 36.870° (b) 53.13° (c) 30.964°

5-4. (a) 86,400 (b) 2.778 (c) 3600 (d) 60 (e) 0.6667 (f) 0.1745 (g) 1.146

5-5. (a) 0.045 (b) 0.022 (c) 0.016 (d) 0.036 (e) 28 (f) 175

5-6. $\sin(10°) = 0.17365 \approx 0.17453$; dev = 0.5%;
$\cos(10°) = 0.98481 \approx 0.98477$; dev = −0.004%
$\tan(10°) = 0.17633 \approx 0.17453$; dev = −1%;
next term corrects to within 1% of deviation in all cases.

5-7. (a) 0.693 (b) 0.301 (c) −0.693 (d) −0.301 (e) 3.162 (f) 0.135 (g) 1.22

5-8. (a) 1 s (b) 0.43 s (c) 0.30 s (d) 0.32

5-9. (a) 47 min (b) 20 min (c) 14 min (d) 0.23

5-10. 44(1) μsec (With RC program, E = 0.1. Answer with linear regression is 42.6(4) μsec)

5-11. (a) 4,800,000 (b) 2,000,000 (c) 200,000

m	0	1	2	3	4	5	6	7
P(m)	0.00	0.01	0.04	0.09	0.13	0.16	0.16	0.14
m	8	9	10	11	12	13	14	
P(m)	0.10	0.07	0.04	0.02	0.01	0.01	0.00	

5-12. 10,560(4) cal/mole K

5-13. 7.00(4) hr (from first 6 data points only).

INDEX

Lichten, Data and Error Analysis, Second Edition
Prentice Hall, Inc.

SYSTEM REQUIREMENTS
The enclosed CD-ROM contains a suite of curve-fitting Excel worksheets which can run on any Macintosh or Windows compatible computer equipped with a CD-ROM and Microsoft Excel version 4 or later (the enhanced FitHaven worksheet requires Excel 5 or later versions and the cT programs will run only on a Mac or Window 95 - compatible machine).
The CD-ROM also contains Macintosh and Windows95 versions of two tutorial programs on error propagation which run in less than 2MB of free RAM.